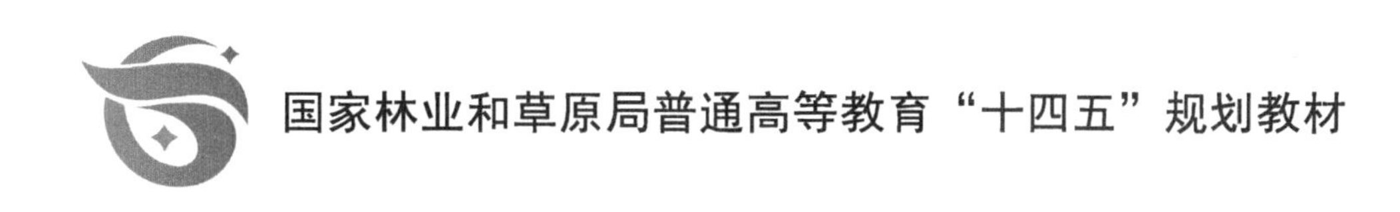

现代林业英语读写教程

Academic English for Modern Forestry: Reading & Writing

李　芝　娄瑞娟　主　编

中国林業出版社
CFPH China Forestry Publishing House

内 容 简 介

《现代林业英语读写教程》的编写理念为“价值引领、学术导向、以读促写、读写并重”。本教程旨在培养高校大学生的学术英语读写能力、逻辑思维能力和批判性思维能力。本教程以学科话题为内容依托，分别从森林和气候变化，森林和生物多样性，森林生态系统，湿地生态系统，草原生态系统，森林国家公园，林业信息技术和林业可持续发展八个学科领域精选学术英语文章。本教程将结构脉络清晰，构思论证严谨，兼具逻辑之美与语言之美作为选文依据，以期从人文角度激发学生对学科研究的思考，从而提高学生的学术思辨能力和学术能力。

图书在版编目（CIP）数据

现代林业英语读写教程 / 李芝, 娄瑞娟主编.
北京 ：中国林业出版社, 2024. 7. (2025.1 重印)-- (国家林业和草原局普通高等教育“十四五”规划教材). -- ISBN 978-7-5219-2760-3

Ⅰ. S7

中国国家版本馆 CIP 数据核字第 2024TS3202 号

策划编辑：高红岩　王奕丹
责任编辑：曹　阳
责任校对：曹　慧
封面设计：睿思视界视觉设计

出版发行：中国林业出版社
（100009，北京市西城区刘海胡同 7 号，电话 83223120　83143611）
电子邮箱：cfphzbs@163.com
网　　址：https://www.cfph.net
印　　刷：北京中科印刷有限公司
版　　次：2024 年 7 月第 1 版
印　　次：2025 年 1 月第 2 次印刷
开　　本：787mm×1092mm　1/16
印　　张：9.5
字　　数：275 千字
定　　价：35.00 元

《现代林业英语读写教程》编写人员

主　　编　李　芝　娄瑞娟

副 主 编　吴增欣　龙　莺　曹荣平

编　　者（按姓氏笔画排序）

龙　莺　刘　真　杜景芬　李　芝

吴增欣　肖　婵　陈声威　郗　佼

娄瑞娟　曹荣平

前　言

Preface

生态文明建设作为国家战略，是推动可持续发展、实现人与自然和谐共生的重要途径。这一国家战略要求我们用行动保护和恢复生态系统，提升公众环保意识和参与度。林业作为维护生态平衡、保护生物多样性的关键行业，其重要性不言而喻。随着全球化进程的深入，国际交流与合作日益频繁，英语成为不同文化、不同专业沟通的桥梁。因此，掌握现代林业英语对于涉林专业的学生和研究人员来说尤为重要。

《现代林业英语读写教程》的编写和出版正是在这样的时代背景之下应运而生的。首先，林业的国际化特征要求林业相关人员不仅要有扎实的专业知识，还要具备良好的英语沟通能力，以便在国际合作和学术交流中发挥作用。其次，随着生态文明理念的普及，公众对于林业和环境保护的关注度不断升高，本教程旨在更有效地传播生态文明建设的理念，讲好中国生态文明建设的故事，进一步提升公众的环保意识，帮助学习者更好地参与有关生态文明建设的国际合作项目，为推动生态文明建设和全球环境治理贡献力量。

《现代林业英语读写教程》的编写秉持“价值引领、学术导向、以读促写、读写并重”的教学理念。内容安排上以阅读促进写作，在广泛阅读学术篇章的基础上，帮助学生掌握学术语篇体裁特征，了解学术论文写作规范，学会批判性阅读，提高学生逻辑思维和思辨能力。本教程选篇结构脉络清晰，构思论证严谨，兼具逻辑之美与语言之美，从而激发学生从人文角度审慎思考科学研究领域的问题，提高学生的思辨能力和学术能力，锻炼学生汲取、处理和传递信息的能力。同时，本教程在信息输入基础上，将语言知识的讲解和语言交际能力的培养结合起来，从而提高读者学术英语写作能力。

本教程在内容选取上，以学科话题为内容依托，分为八个单元，涵盖森林和气候变化，森林和生物多样性，森林生态系统，湿地生态系统，草原生态系统，国家森林公园，林业信息技术和林业可持续发展八个学科领域。每个单元包括两篇学术性文章，长度约600～1200 字。第一篇文章为专业介绍性文章，旨在让学生对该单元的学科领域进行概括性了解，文后练习以阅读技能和阅读理解训练题为主。第二篇文章的学术性特点比较突出，具有学科代表性。文后重点聚焦于学术写作专题的训练。在教学方法上，本教程依据语言和专业知识相融合的理念，以语言产出为导向，打造以内容主题为基础，以阅读思辨能力提升、学术写作能力提升为目标，以读促写、读写训练相结合的构架体系。

综上所述，编写《现代林业英语读写教程》不仅能够为涉林相关专业的学生和研究人员提供一个学习和提高专业英语的机会，帮助他们更好地理解和掌握国际林业领域的最新动态和发展趋势，还能够为生态文明建设贡献自己的力量，为推动全球生态环境的可持续发展尽绵薄之力。

本教程系教育部新文科研究与改革实践项目“产出导向与持续质量改进模式下的新文科外语类课程体系和教材体系建设与实践研究”（项目编号：2021070015）和北京林业大学教育教学改革与研究项目“多层次模块化链条式大学英语人才培养课程体系建设”（BJFU2022JYZD012）的阶段性成果。第一单元由曹荣平编写，第二单元由吴增欣编写，第三单元由陈声威编写，第四单元由杜景芬编写，第五单元由刘真编写，第六单元由龙莺编写，第七单元由肖婵编写，第八单元由郗佼编写，全书李芝、娄瑞娟统稿并校对。

在此，衷心感谢所有为本教程编写和完善提供支持和帮助的同仁们。我们的工作是建立在前人研究和智慧的坚实基础上的，每一页都凝聚着无数林业专家、语言学者和教育工作者的辛勤努力和深刻见解。我们深知，教程的编写难免有疏漏之处。我们诚挚地希望广大读者不吝赐教，提出宝贵的意见和建议。您的每一条反馈都是我们不断进步和完善的动力。我们也期待与国内外的同行们进行更深入的交流与合作，共同携手培养更多具有国际视野和环保意识的林业人才，为推动生态文明建设，贡献我们的智慧和力量。

编　者

2024 年 6 月

目　录

Contents

Unit 1 Forests and Climate Change

Learning Objectives

After learning this unit, you will be able to:

- Know about the importance of forests in addressing climate change.
- Be acquainted with the language features of academic essays.
- Learn two main types of academic writing: exposition and argumentation.

Part I Reading for Academic Purposes

Academic English Is Different from General English

Academic English, oral or written, is a type of a formal discourse which assumes an objective tone. The language is predominantly used for communication in university classrooms and extracurricular professional activities. Much of academic language is discipline specific and involves subject matter thinking. It is featured by a specific writing style and substantiated in a certain disciplinary vocabulary and syntax. It includes the "language of the discipline" (vocabulary and grammar of language associated with the disciplinary contexts) and the "instructional language" used to engage students in learning.

English is the most popular language used in international academic publications. Without exception, academic English is different from general English for daily life communication in lexicology, syntax and content organization. It takes much time and a lot of practice to acquire academic English competence. Most students are not adequately exposed to this language outside university courses. Unless we make academic English explicitly known to students, some students will be excluded from discourses in academic settings of learning and communication, and hence deprived of future opportunities that depend on this language competency.

Reading academic articles can help students get familiar with the style and standards of academic language. Reading extensively can expose them to a variety of subject matters and technical words or expressions. Critical reading will benefit an active mind. In order to qualify as a writer for a topic, one should be well-informed of its subject matter and recent development in research through extensive reading of existing studies or/and communicating with leading scholars of the field.

Apart from differences between academic English and general English, there are cultural and linguistic gaps between the English and Chinese languages apart from vocabularies, which makes English writing even harder to Chinese learners. Reading for academic writing purposes aims at acquainting students with the academic language style, subject matter knowledge and vocabulary, as well as a good logic for writing.

Reading Passage 1

Critical Thinking Questions

Directions: Read the following passage and answer the following questions.

1. What does "NbS" mean?
2. Why do the authors bring "NbS" into discussion?
3. What benefits do NbS bring to us?
4. What is the authors' attitude towards tree planting?
5. What is the right message about NbS to climate change according to the authors?

Getting the Message Right on Nature-based Solutions to Climate Change

Nathalie Seddon, Alison Smith, Pete Smith, et al.

A. The past two years have seen the publication of several major global synthesis reports that collectively paint a **bleak** picture of the current state of the climate and biosphere. Not only are we failing to **stabilize** the climate or stem the tide of biodiversity loss on land and in the sea, but these failures are increasing poverty and inequality across the globe and are severely **undermining** the development gains of the 20th century. There is a growing realization that these challenges are interlinked and cannot be addressed independently. As evidence builds that the natural systems on which we depend are **deteriorating** beyond a point of no return, it is clear that larger scale and more coherent approaches to tackling global challenges are needed.

B. Nature-based solutions (NbS)—solutions to societal challenges that involve working with nature—have recently gained popularity as an **integrated** approach that could address the twin crises of climate change and biodiversity loss, while supporting a wide range of sustainable development goals. NbS are actions that are broadly **categorized** as the protection, restoration or management of natural and semi-natural ecosystems, sustainable management of working lands and aquatic systems, or the creation of novel ecosystems. Although more research is needed, a rapidly growing evidence base demonstrates that well-designed NbS can deliver multiple benefits. For example, protecting and restoring habitats along shorelines or in upper catchments can contribute to climate change adaptation by protecting communities and **infra-**

structure from flooding and erosion, at the same time as increasing carbon sequestration and protecting biodiversity. Meanwhile, increasing green space and planting trees in urban areas can help with cooling and flood **abatement** while mitigating air pollution, providing recreation and health benefits and **sequestering** carbon.

C. Although NbS **span** a wide range of actions, from protection and restoration of **terrestrial** and marine ecosystems to sustainable agriculture and urban green infrastructure funds are currently being channelled mainly towards tree planting. The simple and powerful narrative of "plant a tree to save the planet" is universally appealing, but over-reliance on tree planting as a climate solution raises a number of concerns. First, planting trees does not **equate** to establishing a healthy forest with a complex functional web of interactions among multiple species, which often necessitates careful **stewardship** over many years if not decades. Second, inappropriate tree planting can do more harm than good, especially **afforestation** of naturally open habitats, or planting on high carbon soils. For example, although afforestation increases topsoil carbon in carbon-poor soils, the associated soil disturbance causes significant losses in carbon-rich soils, especially of the more **resilient** deep soil carbon which takes many decades to accumulate. This suggests that the widely used method of estimating soil carbon from a fixed ratio with vegetation biomass overestimates carbon sequestration from afforestation. Afforestation on peaty soils can lead to losses of soil carbon that outweigh that sequestered as the trees grow.

D. Third, afforestation can also reduce ecosystem resilience and thus long-term carbon storage and sequestration. For example, fire-adapted savannah and dryland grassland ecosystems hold large carbon stores below ground. They readily recover from the relatively cool and frequent grassland fires, which do not destroy soil carbon, but afforestation risks much greater carbon losses during intensely hot plantation fires, and can also increase the risk of fires on peatland in **temperate** regions.

E. Fourth, tree-planting schemes must be carefully designed if they are to deliver the intended benefits. For example, mangroves can only thrive in particular conditions of soil, climate, tidal **fluctuations** and wind velocity. **Compensatory** offsets and afforestation schemes that ignore these factors have often resulted in slow and stunted growth. Many investments **badged** as NbS are for commercial plantations, which do not provide permanent carbon stores. Although harvested timber can lock up carbon in long-lived products such as timber-framed buildings or furniture, the carbon stored in these products has been overestimated. A high proportion of harvested wood is used for paper, card and short-lived products such as MDF furniture, which soon end up in landfill or **incineration**, releasing carbon back to the atmosphere so that the net result could even be a carbon loss.

F. Fifth, trees in the wrong place can also cause **trade-offs** between ecosystem services. More research is needed into the **dynamics** of these trade-offs, but current evidence shows that, for example, single-aged, low diversity, intensively managed plantations deliver wood products but may cause water pollution from soil disturbance and agrochemical use and reduce water availability in **arid** regions.

G. Finally, and critically, the current focus on planting trees is **distracting** from the urgent need to effectively protect remaining intact ecosystems. Less than 1% of tropical, temperate and montane grasslands, tropical coniferous forests, tropical dry forests and mangroves are classed as intact, that is, having very low human influence. Not only are these intact ecosystems hotspots for biodiversity, but intact old-growth forests are particularly important for carbon storage and sequestration while also protecting people from climate change impacts. Yet, many of the world's remaining intact ecosystems lack effective protection or are poorly managed; including marine-protected areas where dredging takes place. Degradation of terrestrial habitats (e.g. through logging, drainage, infrastructure development) significantly reduces carbon storage and increases **vulnerability** to climate-related hazards such as fire. Freshwater, coastal and marine habitats face similar issues due to water pollution, temperature increases, sea-level rise, over-fishing, the spread of **invasive** species and, in some cases, inappropriate management. A balanced NbS approach would give greater priority to protecting these remaining **intact** ecosystems, as well as restoring partially degraded forests, and other approaches such as "proforestation"—leaving forests to grow to their full potential, with minimal intervention, and natural **regeneration** of native ecosystems, where appropriate.

H. In summary, a more **holistic** approach is needed which protects, restores and connects a wide range of ecosystems across landscapes and seascapes, including native woodlands, shrublands, savannas, wetlands, grasslands, reefs and seagrass, as well as sustainable agriculture and urban green infrastructure. This will **identify** which ecosystems are appropriate to suit the local ecological and climate context, and will balance local needs for food and materials with the need to support biodiversity, climate change adaptation and other sustainable development goals. To support investment in a diverse range of habitats, we also need to extend current metrics and standards beyond those used for forest carbon to include other carbon-rich habitats such as wetlands and grasslands.

Key Words and Phrases

1. bleak /bliːk/ *adj.* offering little or no hope 不乐观的，无望的
2. stabilize /'steɪbəlaɪz/ *v.* become stable or more stable（使）稳定，稳固
3. undermine /ˌʌndə'maɪn/ *v.* make something such as a feeling or a system less strong or less secure than it was before 逐渐削弱（损害）
4. deteriorate /dɪ'tɪəriəreɪt/ *v.* become worse 恶化，变坏
5. integrate /'ɪntɪgreɪt/ *v.* to combine two or more things so that they work together（使）合并，

加入，融为一体

6. categorize /'kætəgəraɪz/ *v.* divide into or assign to a group 将……分类，归类为

7. infrastructure /'ɪnfrəstrʌktʃə(r)/ *n.* the basic structure or features of a system or organization 基础设施，基础建设

8. abatement /ə'beɪtmənt/ *n.* reduction in the strength or power of something 减少；消除；减轻

9. sequester /sɪ'kwestə(r)/ *vt.* set apart, isolate from or prevented from contact with others 使隔绝；使隐退；没收，扣押（同 sequestrate）

10. span /spæn/ *v.* to cover or extend over an area or time period 持续，贯穿；横跨，涵盖（多项内容）

11. terrestrial /tə'restriəl/ *adj.* of or relating to or inhabiting the land as opposed to the sea or air 地球上的；陆生的；陆地（上）的

12. equate /ɪ'kweɪt/ *v.* consider or describe as similar, equal, or be equivalent or parallel （使）等同；使相等

13. stewardship /'stjuːədʃɪp/ *n.* (formal) the act of taking care of or managing something, for example property, an organization, money or valuable objects 管理；看管；组织工作

14. afforestation /əˌfɒrɪ'steɪʃn/ *n.* (technical) the process of planting areas of land with trees in order to form a forest 植树造林

15. resilient /rɪ'zɪliənt/ *adj.* strong and not easily damaged, returning to its original shape after being bent, stretched, or pressed 有弹性的，有复原力的；对困境有承受力的

16. temperate /'tempərət/ *adj.* (of weather or climate) free from extremes; mild 温带的，（气候）温和的

17. fluctuation /ˌflʌktʃu'eɪʃ(ə)n/ *n.* the quality of being unsteady and subject to changes 波动，起伏

18. compensatory /ˌkɒm.pən'seɪtəri/ *adj.* given or paid to someone in ex-

			change for something that has been lost or damaged, or to pay for something that has been done 赔偿的；弥补的，补偿性的
19.	badge /bædʒ/	*v.*	mark or distinguish (someone or something) with or as if with a badgc 给予……徽章，标记为
20.	incineration /ɪnˌsɪnə'reɪʃ(ə)n/	*n.*	the act of burning something completely 焚化；烧成灰
21.	trade-off /'treɪdˌɒf/	*n.*	the act of balancing two things that you need or want but which are opposed to each other 权衡，协调
22.	dynamics /daɪ'næmɪks/	*n.*	[pl.] the way in which people or things behave and react to each other in a particular situation 动态
23.	arid /'ærɪd/	*adj.*	lacking sufficient water or rainfall （土地或气候）干燥的，干旱的；枯燥的，乏味的
24.	distract /dis'trækt/	*v.*	draw someone's attention away from something 分散，打扰，使分心
25.	vulnerability /ˌvʌlnərə'bɪləti/	*n.*	the state of being vulnerable or exposed the possibility of being attacked or harmed, either physically or emotionally 易损性，弱点
26.	invasive /ɪn'veɪsɪv/	*adj.*	something undesirable that spreads very quickly and that is very difficult to stop from spreading 扩散性的，侵入的
27.	intact /ɪn'tækt/	*adj.*	complete and has not been damaged or changed 完好无损的
28.	regeneration /rɪˌdʒenə'reɪʃn/	*n.*	renewal or restoration of a body, bodily part, or biological system (such as a forest) after injury or as a normal process 再生，重生
29.	holistic /həu'listik/	*adj.*	emphasizing the organic or functional relation between parts and the whole 整体的，全面的
30.	identify/ aɪ'dentɪfaɪ/	*v.*	recognize them or distinguish them from others, determine the taxono-

mic position or category of (a biological specimen) 认出，识别；查明，确认

Reading Comprehension

Directions: Read **Passage 1** again and answer the following questions.

1. Para.1 states __________.

A. the purpose of writing

B. the background of the research topic

C. the significance of the study

D. the framework of the research presentation

2. In Para. 2, "an integrated approach" is closest in meaning to __________.

A. a holistic approach

B. a focused approach

C. an effective approach

D. a related approach

3. What is the right message about "tree-planting schemes" according to the passage?

A. Tree planting as a climate solution is well grounded.

B. The benefits of tree planting will never be over emphasized.

C. Focus on planting trees is inappropriately sucking money.

D. Planting trees is a balanced NbS approach to climate change.

4. The main purpose of the passage is to __________.

A. address the importance of stabilizing the climate

B. state the benefits of NbS

C. raise questions about climate change

D. call attention to a wide range of NbS

Vocabulary Excersies

Exercise 1

Directions: In this section, there are ten sentences, each with a word missing. You are required to complete these sentences with the proper form of the words given in the brackets.

1. This will balance local needs for food and materials with the need to support biodiversity, climate change adaptation and other ____________ development goals. (sustain)

2. A balanced NbS approach would give greater ____________ to protecting these remaining intact ecosystems, as well as restoring partially degraded forests. (prior)

3. These failures are increasing poverty and ____________ across the globe and are severely undermining the development gains of the 20th century. (equal)

4. NbS are solutions to ____________ challenges that involve working with nature. (society)

5. The breakdown of the car ____________ a change in family plans. (necessity)

6. Afforestation can also reduce ecosystem ____________ and thus long-term carbon storage and sequestration. (resilient)

7. Freshwater, coastal and marine habitats face similar issues due to water pollution, temperature increases, sea-level rise, over-fishing, the spread of ____________ species and, in some cases, inappropriate management. (invade)

8. Current evidence shows that single-aged, low diversity, intensively managed plantations deliver wood products but may cause water pollution from soil disturbance and agrochemical use and reduce water ____________ in arid regions. (available)

9. Second, ____________ tree planting can do more harm than good, especially afforestation of naturally open habitats, or planting on high carbon soils. (appropriate)

10. Many investments badged as NbS are for ____________ plantations, which do not provide permanent carbon stores. (commerce)

Exercise 2

Directions: In this section, there are ten sentences with ten blanks. You are required to select one word for each blank from a list of choices given in a word bank. Each choice in the bank is identified by a letter. You may not use any of the words in the bank more than once.

A. addressed	B. categorized	C. distract	D. infrastructure	E. resilient
F. integrated	G. mitigating	H. broadly	I. span	J. sequestration
K. stabilize	L. terrestrial	M. range	N. temperate	O. universally

1. A more holistic approach is needed to protect, restore and connect a wide ____________ of ecosystems across landscapes and seascapes.

2. This suggests that the widely used method of estimating soil carbon from a fixed ratio with vegetation biomass overestimates carbon ____________ from afforestation.

3. The following is a list of all the sources consulted, ____________ by period.

4. Nature-based solutions (NbS) have recently gained popularity as a(n) ____________ approach that could address the twin crises of climate change and biodiversity loss.

5. The store will ____________ seven floors in a mixed-use, retail, hotel and high-rise residential tower.

6. There is a growing realization that these challenges are interlinked and cannot be ____________ independently.

7. Protecting and restoring habitats along shorelines or in upper catchments can contribute to climate change adaptation by protecting communities and ____________ from flooding and erosion.

8. The objective of the treaty is to ____________ greenhouse gas concentrations in the atmosphere at a level that would prevent dangerous anthropogenic interference with the

climate system.

9. Degradation of ___________ habitats significantly reduces carbon storage and increases vulnerability to climate-related hazards such as fire.

10. Drought and floods are projected to become a larger problem in many ___________ and humid regions.

Part II Academic Writing Strategies

Exposition and Argumentation in Academic Writing

We write for different purposes: to tell a story, to describe a scene, to expound a thought or to convince an audience. There is a logic hierarchy of complexity in writing, from expressing (a fact or an opinion), to describing (a picture with details in connection) , to explaining (logical relations), to convincing (with reasons to support a claim).

Exposition and argumentation are two major types of writing tasks that college students are often assigned for academic proposes, e.g. course papers, graduation thesis, research articles to be published for conferences or in academic journals.

Exposition means explaining or expounding, with a purpose to inform people of one's knowledge about a theory, a process, or the nature of an issue, for example, the process of making a machine, the causes of a phenomenon, the planning of a project, the solution of a problem. Exposition may involve writing techniques frequently used for narration or description (e.g. illustration with stories or pictures), but more typical techniques for expository writing are good for explaining processes or relations. Table 1 illustrates some typical writing patterns for exposition.

Table 1 Main Patterns for Exposition

Pattern	Description
Definition	Clarification of a concept/a technical term with a statement or statements about its nature, quality, contents, or function
Illustration	Use of specific and substantial examples to support the ideas of the writer or to explain difficult points or unfamiliar concepts
Process Analysis	Use of step-by-step analysis to analyze and explain how something is done
Problems and Solutions	Identification of a specific problem followed by a discussion that offers a possible solution
Classification/Division	Division is to separate a thing into parts while classification is to organize things which share certain qualities into a category. The former stresses the distinction between things, whereas the latter emphasizes similarities
Comparison and Contrast	Comparison means to compare two seemingly unrelated things to explain their similarities and contrast is to analyze two similar things to reveal their significant differences
Cause and Effect	A list of one or more causes and the resulting effect or effects

Reading Passage 2

Critical Thinking Questions

Directions: Read the following passage and answer the following questions.

1. What is the writing style of the following passage?
2. What are the differences in writing style between the two passages of this unit?
3. How does the writer organize the following passage?

Observed Impacts from Climate Change

IPCC Assessment Report

A. Human-induced climate change, including more frequent and intense extreme events, has caused widespread **adverse** impacts and related losses and damages to nature and people, beyond natural climate **variability.** Some development and adaptation efforts have reduced vulnerability. Across sectors and regions, the most vulnerable people and systems are observed to be disproportionately affected. The rise in weather and climate extremes has led to some **irreversible** impacts as SPM natural and human systems are pushed beyond their ability to adapt.

B. Widespread, **pervasive** impacts to ecosystems, people, settlements, and infrastructure have resulted from observed increases in the frequency and intensity of climate and weather extremes, including hot extremes on land and in the ocean, heavy precipitation events, drought and fire weather. Observed increases in areas burned by wildfires have been attributed to human-induced climate change in some regions. Adverse impacts from tropical cyclones, with related losses and damages, have increased due to sea level rise and the increase in heavy precipitation. Impacts in natural and human systems from slow-onset processes such as ocean acidification, sea level rise or regional decreases in precipitation have also been attributed to human **induced** climate change.

C. Climate change has caused substantial damages, and increasingly irreversible losses, in terrestrial, freshwater and coastal and open ocean marine ecosystems. The extent and magnitude of climate change impacts are larger than estimated in previous assessments. Widespread **deterioration** of ecosystem structure and function, **resilience** and natural adaptive capacity, as well as shifts in seasonal timing have occurred due to climate change, with adverse socioeconomic consequences. Approximately half of the species assessed globally have shifted polewards or, on land, also to higher elevations. Hundreds of local losses of species have been driven by increases in the **magnitude** of heat extremes, as well as mass **mortality** events on land and in the ocean and loss of kelp forests. Some losses are already irreversible, such as the first species extinctions driven by climate change. Other impacts are approaching irreversibility such as the impacts of hydrological changes resulting from the **retreat** of glaciers, or the changes in some mountain and Arctic ecosystems driven by permafrost thaw.

D. Climate change including increases in frequency and intensity of extremes have reduced food and water security, **hindering** efforts to meet Sustainable Development Goals. Although overall agricultural productivity has increased, climate change has slowed this growth over the past 50 years globally, related negative impacts were mainly in mid- and low-latitude regions but positive impacts occurred in some high latitude regions. Ocean warming and ocean acidification have adversely affected food production from shellfish aquaculture and fisheries in some oceanic regions. Increasing weather and climate extreme events have exposed millions of people to acute food insecurity and reduced water security, with the largest impacts observed in many locations and/or communities in Africa, Asia, Central and South America, Small Islands and the Arctic.

E. Climate change has adversely affected physical health of people globally and mental health of people in the assessed regions. Climate change impacts on health are **mediated** through natural and human systems, including economic and social conditions and **disruptions**. In all regions extreme heat events have resulted in human mortality and morbidity. The occurrence of climate-related food-borne and water-borne diseases has increased. The incidence of vector-borne diseases has increased from range expansion and/or increased reproduction of disease vectors. In assessed regions, some mental health challenges are associated with increasing temperatures, trauma from weather and climate extreme events, and loss of livelihoods and culture. Increased exposure to wildfire smoke, atmospheric dust, and aeroallergens have been associated with climate-sensitive cardiovascular and **respiratory** distress. Health services have been disrupted by extreme events such as floods.

F. Overall adverse economic impacts attributable to climate change, including slow-onset and extreme weather events, have been increasingly **identified.** Some positive economic effects have been identified in regions that have benefited from lower energy demand as well as comparative advantages in agricultural markets and tourism. Economic damages from climate change have been detected in climate-exposed sectors, with regional effects to agriculture, forestry, fishery, energy, and tourism, and through outdoor labour productivity. Some extreme weather events, such as tropical cyclones, have reduced economic growth in the short-term. Non-climatic factors including some patterns of settlement, and siting of infrastructure have contributed to the exposure of more **assets** to extreme climate **hazards** increasing the magnitude of the losses. Individual livelihoods have been affected through changes in agricultural productivity, impacts on human health and food security, destruction of homes and infrastructure, and loss of property and income, with adverse effects on gender and social equity.

G. Climate change is contributing to humanitarian crises where climate hazards interact with high vulnerability. Climate and weather extremes are increasingly driving **displacement** in all regions, with Small Island States disproportionately affected. Flood and drought-related acute food insecurity and **malnutrition** have increased in Africa and Central and South America. While non-climatic factors are the dominant drivers of existing intrastate violent conflicts, in some assessed regions extreme weather and climate events have had a small, adverse impact on their length, severity or frequency, but the statistical association is weak. Through displacement and involuntary migration from extreme weather and climate events, climate change has generated and **perpetuated** vulnerability.

Key Words and Phrases

1. adverse /'ædvɜːs/ *adj.* negative and unpleasant; not likely to produce a good result 不利的，有害的；相反的
2. variability /ˌveəriə'bɪləti/ *n.* the quality of being subject to variation 可变性，变化性
3. irreversible /ˌɪrɪ'vɜːsəb(ə)l/ *adj.* that cannot be changed back to what it was before 无法挽回的；不能扭转的
4. pervasive /pə'veɪsɪv/ *adj.* present or felt throughout a place or thing（尤指不好的事物）遍布的；弥漫的
5. induce /in'djuːs, in'duːs/ *v.* cause to do or occur 引起，导致；引诱，诱使
6. deterioration /dɪˌtɪəriə'reɪʃ(ə)n/ *n.* process of changing to an inferior state 恶化
7. resilience /rɪ'zɪliəns/ *n.* the ability of people or things to feel better quickly after something unpleasant, such as shock, injury, etc. 快速恢复的能力；适应力
8. magnitude /'mægnitjuːd/ *n.* the property of relative size or extent (whether large or small) 规模，大小；重要性；震级
9. mortality /mɔː'tæləti/ *n.* a death or the ratio of deaths in an area 死亡，死亡人数，死亡率
10. retreat /rɪ'triːt/ *n.* a movement away from a place 后退，离开；撤兵
11. hinder /'hində/ *v.* to make it more difficult for sb. to do sth. or make progress 阻碍，妨碍
12. mediate /'miːdieit/ *v.* (formal) to influence something and/

		or make it possible for it to happen 影响……的发生；使……可能发生
13. disruption /dis'rʌpʃən/	*n.*	a disorderly outburst or the act of causing disorder 扰乱
14. respiratory /'respərətɔ:ri/	*adj.*	connected with breathing 呼吸的
15. identify /aɪ'dentɪfaɪ/	*v.*	to establish or indicate who or what (someone or something) is 认出，识别；查明，确认
16. asset /'æset/	*n.*	a useful or valuable thing, person, or quality or property owned by a person or company 有价值的人或物；资产，财产
17. hazard /'hæzəd/	*n.*	a source of danger; a possibility of incurring loss or misfortune 危险，危害
18. displacement /dɪs'pleɪsmənt/	*n.*	the removal of something from its usual place or the forcing of people away from the area or country where they live 取代，代替；被迫迁徙，逐出家园
19. malnutrition /ˌmælnju'trɪʃ(ə)n/	*n.*	a state of poor nutrition 营养不良
20. perpetuate /pə'petʃueit/	*v.*	make (something, typically an undesirable situation or an unfounded belief) continue indefinitely 使持续；使持久

Writing Tasks

Exercise 1

Directions: Read the following sentences from **Passage 2** and rewrite the following sentences (underlined in the passage), replacing the underlined parts with your own words.

1. Some losses are already irreversible, such as the first species extinctions driven by climate change.

2. Overall adverse economic impacts attributable to climate change, including slow-onset and extreme weather events, have been increasingly identified.

3. Non-climatic factors including some patterns of settlement, and siting of infrastructure have contributed to the exposure of more assets to extreme climate hazards increasing the magnitude of the losses.

4. Through displacement and involuntary migration from extreme weather and climate events, climate change has generated and perpetuated vulnerability.

Exercise 2

Directions: Read **Passage 2** again and answer the questions below. Please give brief answers in about 10 words.

1. What caused the increases in areas burned by wildfires in some regions? (Para. B)
2. How many species have left their original habitats to polewards or, on land, also to higher elevations, based on the global assement? (Para. C)
3. Why didn't the increase of agricultural production contribute to the rise of the global agricultural production over the past 50 years? (Para. D)
4. How does climate change exert impacts on health? (Para. E)
5. In what regions does climate change have some positive economic effects identified? (Para. F)

Exercise 3

Directions: Read the passage again and write a summary of **Passage 2** with no more than 150 words.

Extensive Reading and Writing

Directions: For this part, you are required to read the following passage and then write an essay on **Addressing Climate Change**. You should write at least 120 words but no more than 180 words.

Forests and Climate Change: From Complex Problem to Integrated Solution

Hans Hoogeveen

Global warming has become everyday news, often featured in alarming statements by heads of governments, scientists or environmental activists. We now know that melting glaciers, erratic global weather patterns, droughts, raging wildfires and creeping invasive species of flora and fauna in new localities are all unmistakably the effects of climate change.

Skeptics of global warming argue that changes in weather patterns are part of the natural variability in the Earth's temperature, but the majority of scientists agree they are most likely due to human-induced increased concentration of heat-trapping greenhouse gases (GHG) in the atmosphere.

It is crucial for mankind to accept the fact that there is no way to ignore the signs of danger and the risks of the looming global climate change. There is no time to spare, we must act now. This is an excellent opportunity for all stakeholders to meet this challenge through a comprehensive approach in addressing the man-made causes of global warming, in order to create a better and brighter future worthy of the next generations. Scientists have long understood the role forests play in creating microclimates. With increasing awareness on global warming and its main culprit, carbon dioxide (CO_2) emissions, the role of forests and plant resources in modifying the impacts of climate change is gaining renewed attention of climatologists, foresters, policymakers and the media worldwide.

The Fourth Assessment Report of the Intergovernmental Panel on Climate Change (IPCC) has reconfirmed that the increasing GHG emissions due to human activities have led to a marked increase in atmospheric GHG concentrations. Between 1970 and 2004, global GHG emissions have increased by 70 percent; CO_2 emissions alone have grown by about 80 percent (28% between 1990 and 2004) and represented 77 percent of total anthropogenic GHG emissions in 2004. While the largest growth in global emissions from 1970—2004 came from the energy supply sector (an increase of 145%), growth from other sectors was also significant. Emissions from transport, industry, and land use, land-use change and forestry sectors were 120, 65 and 40 percent, respectively.

Although the facts and figures are clear and known, the question remains: What are governments and other stakeholders willing to do to address global warming? Is the international community really committed "to come out of the woods" and bring coherence to its approach in going beyond the strict mandates and competences of the international forest-related processes?

Trees and forests remove carbon dioxide from the atmosphere through photosynthesis to carbon, and store carbon in the form of wood and vegetation—a process referred to as "carbon sequestration". Trees are generally about 20 percent carbon by weight. In addition, the overall biomass of forests also acts as a "carbon sink". Thus, increasing storage and preventing stored carbon from releasing back to the atmosphere are important measures for combating global warming and conserving the environment.

Forests are intricately linked to climate change, both as a cause and a solution. Global climate changes impact the health, distribution and composition of forests. There is increasing evidence that forests are under pressure. Therefore, integrated action should be taken to manage these complex relationships. With the proliferation of international environmental institutions within the United Nations system, the role of forests in mitigating the adverse impacts of climate change is increasingly being addressed in a variety of policy arenas. It is becoming clear that institutional fragmentation leads to incoherence and duplication.

With regard to climate change and forests, the United Nations Framework Convention on Climate Change (UNFCCC), with its specific mandate to combat global warming, is an obvious institutional locus. Another, perhaps lesser-known, body is the United Nations Forum on Forests (UNFF). It is a challenge for both organizations to join hands in making bold leaps forward. UNFCCC and its Kyoto Protocol provided a broader framework to address climate change challenges with specific emission reduction targets, obligations and mechanisms.

In addition to the important developments within the UNFCCC framework, the seventh session of the UNFF in April 2007 adopted a nonlegally binding instrument on all types of forests. After the adoption of the Forest Principles at the 1992 UN Conference on Environment and Development, the international community demonstrated leadership in adding a robust new chapter on the global forest policy, which supports actions on the ground. The linkage between forests and climate change was not only identified by UNFF as highly pertinent, but also observed in a broader, more holistic approach.

In dealing with climate change and forests, the following issues call for immediate and medium-term attention, in order to make a positive contribution of forests to mitigate climate change, adapt forest management to the changing climatic condition, and safeguard the benefits and interests of stakeholders.

Work synergistically and collaboratively. UNFCCC and UNFF should work collaboratively on forest-related climate change issues, as both cannot achieve their objectives on their own. The Collaborative Partnership on Forests (CPF), formed to support the work of UNFF, provides a pathway for such collaboration. Both UNFCCC and UNFF Secretariats, together with 12 other forest-related international organizations, instrument secretariat and institutions, should join forces to seek linkages reaching beyond traditionally demarcated competences and lines of operation.

More coherence within the UN system. Member States operating within the different governing bodies on international forest policies and climate change should convey consistent messages to relevant bodies. Time and again, lack of internal coordination at the country level results in incoherent, and sometimes conflicting, political signals. Robust and forward-looking decisions can be made in shaping the future agenda, only when Member States speak with one voice.

See the bigger picture. Looking at forests for climate change mitigation must take into consideration sustainable development, poverty eradication, rights of indigenous and local communities to forest resources, conservation of biodiversity and other environmental benefits of forests, such as air and water.

Prevent deforestation. Avoid perverse incentives to deforest and provide economic incentives to prevent deforestation, as well as establish afforestation and reforestation projects.

Carbon accounting. Methodological issues related to carbon accounting, including the development of criteria and indicators, and the inherent problems of additionality, leakage and permanence, should be addressed as early as possible.

Strengthen legal instruments. In response to the issues identified above, take advantage of the recently adopted nonlegally binding instrument on all types of forests and the UNFF multi-year programme of work to develop and implement a common policy base on the issue of forests, focusing action on the ground.

While forests show significant societal and environmental potential, the main players, including Governments, business and industry, are in severe need of increased means of implementation, including financial resources, capacity-building and technology transfer. To achieve effective solutions to alarming rates of deforestation and forest degradation, as well as mitigate climate change, the international community as a whole needs to pool resources and share both knowledge and financial resources.

Self-assessment

Read the self-assessment guidelines for *Self-assessment scale for reading comprehension* and *Self-assessment scale for written expression* (based on the *China's Standards of English Language Ability*) outlined in the appendix of this textbook. Please conduct a self-assessment to evaluate your own skills respectively.

Unit 2 Forest and Biodiversity

Learning Objectives

After learning this unit, you will be able to:

- Know about some affecting factors of biodiversity loss.
- Synthesize ideas from sources.
- Develop some writing skills for cause and effect essay.

Part I Reading for Academic Purposes

Argumentative Essays

An argumentative essay is a genre of writing that takes a stance on an issue. The purpose of the essay is to persuade readers into understanding and supporting their point of view about a topic by reasoning.

An argumentative essay contains three core elements (Table 2): claim (thesis), evidence (grounds), reasoning (warrant). A thesis is the assertion that authors would like to prove to their audience. It is the main argument. The grounds of an argument are the evidence and facts that help support the claim. The reasoning (warrant), either implied or stated explicitly, is the assumption that links the evidence to the claim. In other words, the reasoning element explains why the evidence can support the claim.

Table 2 Basic structure of an argumentative essay

Basic structure	Details
Introduction	✧ The thesis must be arguable. To be arguable, a thesis must have some probability of being true and it must be a statement with which people may disagree. ✧ A thesis contains both an observation and an opinion
Body	✧ Present clear and convincing evidence. Support your thesis statement with different forms of evidence, such as facts, statistics, authorities, anecdotes etc.. The evidence can be gathered from various sources. It is important to document the evidence properly. ✧ Present a reasoning to explain how and why the evidence supports the thesis. Do not assume that the evidence will speak for itself. It is essential to explain the importance of each piece of evidence, that is, how it supports your point, why it is significant. ✧ Counterarguments including objections, alternatives, challenges, or questions to the argument may also be anticipated and addressed in some essays. A writer can address counterarguments by acknowledging, accommodating, and/or refuting them

continued table

Basic structure	Details
Conclusion	✧ Restate the topic and the thesis. ✧ Synthesize the main points in the body part. ✧ Do not introduce any new information to the final section

Reading Passage 1

Critical Thinking Questions

Directions: Read the following passage and answer the following questions.

1. How do you understand the phrase "biodiversity hotspots" in the title?
2. Why is biodiversity on islands more vulnerable to global climate change?
3. What measures should be taken to address the issue of biodiversity loss on small islands?

Climate Change and Terrestrial Biodiversity Hotspots on Small Islands

Costello, M. J., M. M. Vale, et al.

A. Despite covering approximately 2% of the Earth's land area, islands **harbour** more than 20% of **extant** terrestrial species. Islands have disproportionately higher rates of **endemism** and threat when compared to continents, with 80% of historical extinctions (since 1500 CE) having occurred on islands. Current climate change projections suggest that **insular** species are particularly sensitive and, even at mild warming levels, **substantial** losses are expected. Given islands' characteristic high endemicity, current high threat levels and the fact that islands host almost half of all species currently considered to be at risk of extinction, especially at higher warming levels, further losses could contribute disproportionately to global biodiversity decline.

B. The high **vulnerability** of terrestrial biodiversity on islands to global change can be explained by a number of limitations, characteristic of both islands and insular species. Older, isolated islands tend to have fewer species and lower functional **redundancy** but a higher proportion of endemism. Many of these islands contain species with inherently high sensitivity to environmental change (narrow habitat ranges, small population sizes, low genetic diversity and poor adaptive, dispersal and defensive capabilities). Unlike continental environments, insular species often have limited opportunities for autonomous adaptation from not having enough geographic space to shift their ranges to track suitable climatic conditions. Local extinction risks are amplified by even small losses of habitat due to global change including human-induced disturbances, extreme events, sea level rise and invasive species.

C. However, some insular species have shown resilience to climate change. Intact island forests, for example, have shown rapid recovery rates after tropical **cyclones**, despite high levels of initial damage, especially in the **Caribbean**. Additionally, many **Mediterranean** islands are "disturbance adapted", with continued persistence of some single-island endemic plants, despite exposure to multiple threats. This continued persistence has been attributed, at least partially, to climate refugia, oceanic **buffering** and high habitat **heterogeneity** within **topographically** complex mountainous regions. However, this climate resilience will not be sustained under climate change, especially when coupled with habitat **degradation**.

D. Adaptation strategies depend on the ability to project future impacts from climate change, but this is **hampered** by lack of fine-scale climate data, especially for developing small island nations. There is a **paucity** of robust impacts-based modelling output for terrestrial biodiversity from these islands due to the wide, **chronic** unavailability of Regional Climate Model (RCM) data **premised** on the most recent suite of **scenarios** (RCPs and especially SSPs). Additionally, realistic assessments of changing climate on such small ecosystems require further RCM downscaling and verification to sub-island resolutions of <5 km. Furthermore, widely used statistically downscaled data at sub-5 km resolutions, such as WorldClim are often unsuitable due to limited spatial and **temporal** resolutions of observation station data from small islands and higher errors associated with statistical downscaling and locations with complex topography and coastlines. Widespread unavailability of such data constrains accurate simulations of climatic variation within the small-scale mountainous and coastal regions of islands, associated with climate refugia and high habitat heterogeneity. This is a key element contributing to the continued delay in development of robust adaptation strategies towards not only biodiversity conservation but other important cross-sectoral issues.

E. Due to islands' limited size and isolation, conventional conservation measures focused on expanding protected areas, **dispersal** corridors and buffer zones are of limited effectiveness on islands (high confidence). Instead, multifaceted, locally driven holistic climate-smart strategies across mosaics of human-impacted, often heavily degraded and **fragmented**, landscapes are required. These should ideally be long-term, flexible and sustainable solutions that **incorporate** social and biocultural knowledge as well as economic co-benefits to island communities in order to "buy time". Examples include ecosystem-based approaches such as ridge-to-reef management, which incorporates conservation partnerships among lands inside and outside protected areas to increase connectivity and reduce land use impacts, while building on the interconnections among terrestrial, freshwater, coastal and marine ecosystems. Such strategies require raising awareness of biodiversity values among local communities, and cross-sectoral planning and policy at both island, regional and trans-boundary scales. These lend to private-public partnerships, increasing the potential of solutions reaching beyond protected areas boundaries and affecting sociopolitical change.

F. Limited **terrain**, natural, economic and data resources across small developing nation islands mean that unconstrained habitat destruction and degradation cannot be sustained, as this harms both people and the biodiversity upon which they depend. This limitation of resources compromises climate adaptation, which is often further complicated by varying **governance** and states of economic development. With changing climate conditions, there is an increased urgency to rethink how progress can be measured, and to create opportunities building on **synergies** between disaster risk reduction, food security and social justice, so that islands can most benefit from their natural resources and biodiversity in a sustained manner.

Key Words and Phrases

1. terrestrial /tə'restriəl/	*adj.*	relating to the earth 地球上的，陆栖的
2. hotspot /'hɒtspɒt/	*n.*	a place where war or other fighting is likely to happen 多事之地，热点地区
3. harbour /'hɑːbə(r)/	*v.*	to contain something and allow it to develop 庇护，带有（为……提供住所或动物栖息地）
4. extant /ek'stænt/	*adj.*	used to refer to something very old that is still existing 现存的
5. endemism /'endɪmɪzəm/	*n.*	nativeness by virtue or originating or occurring naturally (as in a particular place) 特有现象，地方特殊性
6. insular /'ɪnsjələ(r)/	*adj.*	relating to or characteristic of or situated on an island 海岛的
7. substantial /səb'stænʃ(ə)l/	*adj.*	large in amount, value or importance 大量的；价值巨大的
8. vulnerability /ˌvʌlnərə'bɪləti/	*n.*	the fact of being weak and easily hurt physically or emotionally 易损性，弱点
9. redundancy /rɪ'dʌndənsi/	*n.*	the state of not being necessary or useful 多余，冗余
10. cyclone /'saɪkləʊn/	*n.*	a violent tropical storm in which strong winds move in a circle 气旋，龙卷风
11. Caribbean /ˌkærɪ'biːən/	*n.*	the region consisting of the Caribbean Sea and its islands, including the West Indies, and the coasts that surround it 加勒比海，加勒比地区
12. Mediterranean /ˌmedɪtə'reɪniən/	*adj.*	the largest inland sea between Europe

		and Africa and Asia; relating to the Mediterranean Sea or the countries that surround it 地中海地区；地中海的
13. buffer /'bʌfə(r)/	*v.*	to provide protection against harm 保护，缓冲
14. heterogeneity /ˌhetərədʒə'niːəti/	*n.*	the quality of being unsteady and subject to changes 异质性，非均匀性
15. topographically /ˌtɒpə'græfikli/	*adv.*	in a way that is connected with the physical features of an area of land, especially the position of its rivers, mountains, etc. 在地形构造方面
16. degradation /ˌdegrə'deɪʃn/	*n.*	the process of something being damaged or made worse 损害，衰退
17. hamper /'hæmpə(r)/	*v.*	to prevent someone doing something easily 阻碍，妨碍
18. paucity /'pɔːsəti/	*n.*	a small amount of something; less than enough of something 缺乏，少量
19. chronic /'krɒnɪk/	*adj.*	(of a disease; of a problem) lasting a long time（疾病）慢性的，长期的；（问题）难以根除的
20. premise /'premɪs/	*v.*	set forth beforehand, often as an explanation 以……为基础
21. scenario /sə'nɑːriəʊ/	*n.*	a postulated sequence of possible events 可能发生的情况
22. temporal /'tempərəl/	*adj.*	of or relating to or limited by time 时间的
23. dispersal /dɪ'spɜːs(ə)l/	*n.*	the act of dispersing or diffusing something 传播，散布，疏散
24. fragment /'frægmənt/	*v.*	break or cause to break into pieces（使）破碎，分裂
25. incorporate /ɪn'kɔːpəreɪt/	*v.*	make into a whole or make part of a whole 包含，合并，混合（成分）
26. terrain /tə'reɪn/	*n.*	a piece of ground having specific characteristics or military potential 地形，地势，领域
27. governance /'gʌvənəns/	*n.*	the act of governing; exercising authority 统治方式，管理方法
28. synergy /'sɪnədʒi/	*n.*	the working together of two things (muscles or drugs for example) to produce an effect greater than the sum of their individual effects 协同增效作用，协同作用

Reading Comprehension

Directions: Read **Passage 1** again and answer the following questions.

1. __________ is a correct statement according to Para. 1.

A. The species amount on islands is relatively small compared with other types of ecosystems.

B. The amount of endangered species on islands is relatively small.

C. The situation of biodiversity destruction on islands has undergone a great change.

D. The situation of species loss on islands is grimmer compared with continental environments.

2. According to Para. 2, __________ is NOT the reason for the losses of habitat.

A. alien species

B. afforestation

C. the rise of sea level

D. destruction caused by humans

3. __________ is the main obstacle to adaptation strategies.

A. Limited spatial and temporal resolution

B. Human activity

C. Unavailability of relevant data

D. Complex topography and coastlines

4. According to the passage, the measures to address the issue should be __________.

A. from being climate vulnerable to being climate smart

B. long-term, flexible and sustainable

C. beneficial to both economy and society

D. all above

Vocabulary Exercises

Exercise 1

Directions: In this section, there are ten sentences, each one with one word missing. You are required to complete these sentences with the proper form of the words given in the brackets.

1. Islands have ____________ higher rates of endemism and threat when compared to continents, with 80% of historical extinctions (since 1500 CE) having occurred on islands (Taylor and Kumar, 2016; Spatz et al., 2017; Dueñas et al., 2021). (proportion)

2. Intact island forests, for example, have shown rapid ____________ rates after tropical cyclones, despite high levels of initial damage. (recover)

3. This climate resilience will not be sustained under climate change, especially when coupled with habitat____________. (degrade)

4. Many of these islands contain species with inherently high ____________ to en-

vironmental change. (sensitive)

5. Societies need to develop highly ____________ behavioral rules for survival. (adapt)

6. Any sleep ____________ will diminish mental performance. (disturb)

7. The manager prepared a computer ____________ of likely sales performance for the rest of the year. (simulate)

8. Researchers believe wide ____________ of cheap food is one of the main drivers. (available)

9. By most measures, corporate ____________ has become a lot tighter and more rigorous since the 1970s. (govern)

10. It was once assumed that improvements in telecommunications would lead to more ____________ in the population as people were no longer forced into cities. (disperse)

Exercise 2

Directions: In this section, there are ten sentences with ten blanks. You are required to select one word for each blank from a list of choices given in a word bank. Each choice in the bank is identified by a letter. You may not use any of the words in the bank more than once.

A. harbour	B. vulnerabilities	C. redundancy	D. premises	E. buffer
F. substantial	G. hamper	H. chronic	I. scenario	J. fragmented
K. governance	L. extant	M. paucity	N. temporal	O. incorporate

1. The findings show a ____________ difference between the opinions of men and women.

2. Those who organized the attacks exploited ____________ in the nation's defenses.

3. I bought a house as a ____________ against inflation.

4. Some greens worry that natural gas could ____________ the development of renewable energy.

5. Most importantly, perhaps, how can I ____________ music into a curriculum that marginalizes the arts?

6. Indigenous civilizations were based on advanced agriculture and well-developed systems of ____________.

7. The upshot is that physics had two theories, based on conflicting____________, that produced the same results.

8. If this weren't such a dire ____________, Moss would have broken out into laughter.

9. The Americas were more ____________ by areas unsuitable for food production or for dense human populations.

10. Today, there are two ____________ copies: one in the British Library, the other in the Folger Shakespeare Library.

Exercise 3

Directions: Read the passage again and translate the following sentences from **Passage 1** (underlined in the passage) into Chinese.

1. Given islands' characteristic high endemicity, current high threat levels and the fact that islands host almost half of all species currently considered to be at risk of extinction, especially at higher warming levels, further losses could contribute disproportionately to global biodiversity decline.

2. These should ideally be long-term, flexible and sustainable solutions that incorporate social and biocultural knowledge as well as economic co-benefits to island communities in order to "buy time".

3. With changing climate conditions, there is an increased urgency to rethink how progress can be measured, and to create opportunities building on synergies between disaster risk reduction, food security and social justice, so that islands can most benefit from their natural resources and biodiversity in a sustained manner.

Part II Academic Writing Strategies

Cause and Effect Essays in Academic Writing

Cause and effect essays (also called reason and result essay) are quite common in academic writing. They analyze the relationship between causes and effects of a particular event or phenomenon. It usually answers the questions "why?" (cause) and "what is the result?" (effect).

Having a clear structure is essential for a successful essay. The outline of cause and effect essay consists of a minimum of four sections—an introduction, at least two body paragraphs, and a conclusion. The two basic ways of a cause-effect essay are called a block pattern and a chain pattern, as shown in the Table 3 and 4.

Table 3 The block pattern

Pattern	Description
Introduction	Introduce your topic and give some background information
Body	Causes: introduce the causes and provide evidence to back up the argument Effects: discuss the effects that come out of the causes mentioned above
Conclusion	Reiterate your thesis statement

Table 4 The chain pattern

Pattern	Description
Introduction	The same as Type 1
Body	Cause 1—effect 1 Cause 2—effect 2 Every cause will be immediately followed by its effect
Conclusion	The same as Type 1

There are many ways to express cause and effect relationships. You can use verbs, or verbal phrases such as *lead to, facilitate, contribute to, create, produce, cause, affect*; you can also use conjunctions, such as *because of, due to, owing to, as a result of.*

Either the cause or effect can come first depending on the voice of the verb (whether the verb is active or passive) or the function of the conjunction, as shown in the following examples:

(1a) The weakness is caused by a deficiency of Vitamin C.

(1b) The deficiency of Vitamin C causes the weakness.

(2a) The deficiency of red corpuscles is due to a lack of iron.

(2b) Because of the lack of iron, there is a deficiency of red corpuscles.

Reading Passage 2

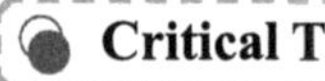

Critical Thinking Questions

Directions: Read the following passage and answer the following questions.

1. What does "cheaper food" paradigm refer to?
2. What is the widely accepted core belief regarding food production?
3. What is the typical way to enhance economic growth through productivity improvements?

How Today's Food System Drives Biodiversity Loss?

Tim G. Benton, Carling Bieg, Helen Harwatt, et al.

A. The role of our food system as the principal driver of biodiversity loss has been shaped by decades of economic growth that has in part been supported by, and in part **incentivizes**, increased food production. This trend also reflects a lack of consideration of the true costs of food production. The drive for increased productivity, and failure to account for the impacts of food production on natural ecosystems and human health, have created and sustained vicious circles that make up what we describe as the "cheaper food" **paradigm**.

Vicious circles in our food system

B. Investment in agricultural productivity, coupled with increased economic competition through the **liberalization** of trade, has long been considered central to a functioning food system. For many, food production is a natural and necessary use of land: people need food, and they depend on the use of land to produce it. Similarly, efforts to reduce food prices are often **deemed** both desirable and necessary: lower food prices deliver two nominal public goods, in the form of increased access to food and reduced household **expenditure** on food.

C. These key **tenets**—that we must produce more food and do so at lower cost if we are to support the global population and drive economic growth—have taken **primacy** over the goals of delivering human and **planetary** health and well-being, with increasingly problematic side-effects. While it is possible to boost economic growth through productivity improvements, this has typically relied on "externalizing" the costs of such improvements on to the environment. In other words, the costs of environmental degradation resulting from food production have not been accounted for and included in the cost of food. Financial incentives such as agricultural **subsidies** are channelled into the food system to increase yields, and the resulting environmental costs—such as pollution through unsustainable production practices—are **discounted** or ignored by the market.

D. As mentioned, reducing food prices through increased productivity can stimulate growth in consumer spending, since it increases the amount of **disposable** income available to buy other goods and services. It also allows consumers to buy more food. Either way, this leads to negative consequences from a planetary health viewpoint: the more disposable income we have, the more we can purchase; the more we can purchase, the more we consume; the more we consume, the more resources we **exploit**; and the more resources we exploit, the more we drive environmental degradation and disrupt natural ecosystems.

The "cheaper food" paradigm

E. Underpinning the "cheaper food" paradigm is a two-way relationship between supply and demand. On the one hand, demand can be seen to shape supply: as the so-called "nutrition transition" around the world has shown, rising incomes tend to **prompt** greater consumption of resource-intensive foods such as animal products, vegetable oils and processed goods, and relatively lower consumption of staple grains. But demand for food—what we eat, how much we eat, and what we waste—is just as much shaped by its supply and price. The more we produce, the cheaper food becomes, and the more we consume. Demand therefore does not simply determine what food is grown and how. It can also be understood as a function of increased supply of cheaper food, and of the way food is processed, marketed and sold. Understanding this relationship between supply and demand is critical to understanding how the current food system drives biodiversity loss, and to identifying effective levers for moving towards a system that supports biodiversity protection and other components of planetary health.

F. The "cheaper food" paradigm drives a set of **overlapping** and often self-reinforcing mechanisms, in which the ratcheting up of production and liberalization of global markets incentivize economic behaviour that creates negative outcomes for society and the environment. These **mechanisms** include the following:

— A drive towards globally competitive markets incentivizes land use for food production at increasing intensity and scale, because the financial rewards are high. The global production system is based on comparative advantage, and thus specialization, with the result that global calorie production is concentrated around a limited set of commodity crops grown using highly intensive methods in a small number of breadbasket regions.

— Intensive farming has a range of negative consequences for the health and quality of soils, air, water sources and natural ecosystems. Partly, this arises from the use of inputs such as **pesticides** and nutrients, and partly it is a function of the **prevalence** of "monocultural landscapes" in which there is little opportunity for nature. In turn, the loss of biodiversity and soil **fertility** leads to a need to intensify agriculture further.

— The concentration and intensification of agriculture have driven down the cost of **staples** such as grains, which are now sufficiently cheap to be diverted from direct human consumption to farmed animals. This has led to growth in the global herd of farmed animals, with negative consequences for air and water quality and GHG emissions.

— As food prices have fallen, it has become increasingly economically rational to waste food. Waste is now occurring at scale along supply chains, creating additional sources of pollution and resulting in "leakage" of the finite resources—including land, water and soil—involved in food production. The more prices fall, the more food we demand and the more we waste; and the more food we waste, the more we demand.

Key Words and Phrases

1. incentivize /ɪn'sentɪvaɪz/	*v.*	to provide (someone) with a good reason for wanting to do something 激励
2. paradigm /'pærədaɪm/	*n.*	a typical example or pattern of something 范例，样板
3. vicious /'vɪʃəs/	*adj.*	the quality of being subject to variation 凶险的，有害的
4. liberalization /ˌlɪbrəlaɪ'zeɪʃn/	*n.*	the act of making less strict 自由化，放宽限制
5. deem /diːm/	*v.*	to consider or judge something in a particular way 认为
6. expenditure /ɪk'spendɪtʃə(r)/	*n.*	the act of spending money for goods or services 经费，支出
7. tenet /'tenɪt/	*n.*	one of the principles on which a belief or theory is based 原则，信条
8. primacy /'praɪməsi/	*n.*	the state of being the most important thing 第一位，首位
9. planetary /'plænətri/	*adj.*	relating to planets 行星的
10. subsidy /'sʌbsədi/	*n.*	money given as part of the cost of something, to help or encourage it to happen 补贴，津贴，补助金
11. discount /'dɪskaʊnt/	*n.*	a reduction in the usual price 折扣，减价
12. disposable /dɪ'spəʊzəb(ə)l/	*adj.*	free or available for use or disposition 可支配的，可自由使用的
13. exploit /ɪk'splɔɪt/	*v.*	use or manipulate to one's advantage 利用，运用

14. prompt /prɒmpt/	*v.*	to cause something to happen 促使，导致
15. overlap /ˌəʊvəˈlæp/	*v.*	coincide partially or wholly（与……）互搭，（与……）复叠
16. mechanism /ˈmekənɪzəm/	*n.*	the technical aspects of doing something 途径，方法
17. pesticide /ˈpestɪsaɪd/	*n.*	a chemical used to kill pests (as rodents or insects) 杀虫剂
18. prevalence /ˈprevələns/	*n.*	the fact of existing or being very common at a particular time or in a particular place 流行，盛行
19. fertility /fəˈtɪləti/	*n.*	the quality in land or soil of making plants grow well 肥沃
20. staple /ˈsteɪp(ə)l/	*n.*	a necessary commodity for which demand is constant 支柱产品，主食

Writing Tasks

Exercise 1

Directions: Read the following sentences from **Passage 2** and rewrite the following sentences (underlined in the passage), replacing the underlined parts with your own words.

1. In other words, the costs of environmental degradation resulting from food production have not been accounted for and included in the cost of food.

2. As mentioned, reducing food prices through increased productivity can stimulate growth in consumer spending, since it increases the amount of disposable income available to buy other goods and services.

3. Either way, this leads to negative consequences from a planetary health viewpoint: the more disposable income we have, the more we can purchase; the more we can purchase, the more we consume; the more we consume, the more resources we exploit; and the more resources we exploit, the more we drive environmental degradation and disrupt natural ecosystems.

4. The "cheaper food" paradigm drives a set of overlapping and often self-reinforcing mechanisms, in which the ratcheting up of production and liberalization of global markets incentivize economic behaviour that creates negative outcomes for society and the environment.

Exercise 2

Directions: Read **Passage 2** again and answer the questions below. Please give brief answers in about 10 words.

1. Why is the reduction of food price generally expected in the whole society? (Para. B)

2. What is the negative impact of increasing food production? (Para. C)

3. What is the significance of understanding the relationship between supply and demand? (Para. E)

4. What is the relationship between food price and waste? (Para. F)
5. What is the passage mainly about? (The whole passage)

Exercise 3

Directions: Read the passage again and write a summary of **Passage 2** with no more than 150 words.

Extensive Reading and Writing

Directions: For this part, you are required to read the following passage and then write an essay on **Addressing Climate Change**. You should write at least 120 words but no more than 180 words.

Biodiversity Loss and Climate Extremes—Study the Feedbacks

Mel D. Mahecha, Ana Bastos, Friedrich J. Bohn, et al.

As humans warm the planet, biodiversity is plummeting. These two global crises are connected in multiple ways. But the details of the intricate feedback loops between biodiversity decline and climate change are astonishingly under-studied.

It is well known that climate extremes such as droughts and heatwaves can have devastating impacts on ecosystems and, in turn, that degraded ecosystems have a reduced capacity to protect humanity against the social and physical impacts of such events. Yet only a few such relationships have been probed in detail. Even less well known is whether biodiversity-depleted ecosystems will also have a negative effect on climate, provoking or exacerbating weather extremes.

For us, a group of researchers living and working mainly in Central Europe, the wake-up call was the sequence of heatwaves of 2018, 2019 and 2022. It felt unreal to watch a floodplain forest suffer drought stress in Leipzig, Germany. Across Germany, more than 380,000 hectares of trees have now been damaged, and the forestry sector is struggling with how to plan restoration activities over the coming decades. What could have protected these ecosystems against such extremes? And how will the resultant damage further impact our climate?

In June 2021, the Intergovernmental Panel on Climate Change (IPCC) and the Intergovernmental Science-Policy Platform on Biodiversity and Ecosystem Services (IPBES) published their first joint report, acknowledging the need for more collaborative work between these two domains. And some good policy moves are afoot: the new EU Forest Strategy for 2030, released in July 2021, and other high-level policy initiatives by the European Commission, formally recognize the multifunctional value of forests, including their role in regulating atmospheric processes and climate. But much more remains to be done.

To thoroughly quantify the risk that lies ahead, ecologists, climate scientists, remote-sensing experts, modelers and data scientists need to work together. The upcoming meeting of the United Nations Convention on Biological Diversity in Montreal, Canada, in December is a good opportunity to catalyse such collaboration.

Buffers and responses

When lamenting the decline in biodiversity, most people think first about the tragedy of species driven to extinction. There are more subtle changes under way, too.

For instance, a study across Germany showed that over the past century, most plant species have declined in cover, with only a few increasing in abundance. Also affected is species functionality—genetic diversity, and the diversity of form and structure that can make communities more or less efficient at taking up nutrients, resisting heat or surviving pathogen attacks.

When entire ecosystems are transformed, their functionality is often degraded. They are left with less capacity to absorb pollution, store carbon dioxide, soak up water, regulate temperature and support vital functions for other organisms, including humans. Conversely, higher levels of functional biodiversity increase the odds of an ecosystem coping with unexpected events, including climate extremes. This is known as the insurance effect.

The effect is well documented in field experiments and modelling studies. And there is mounting evidence of it in ecosystem responses to natural events. A global synthesis of various drought conditions showed, for instance, that forests were more resilient when trees with a greater diversity of strategies for using and transporting water lived together.

However, biodiversity cannot protect all ecosystems against all kinds of impacts. In a study this year across plots in the United States and Canada, for example, mortality was shown to be higher in diverse forest ecosystems. The proposed explanation for this unexpected result was that greater biodiversity could also foster more competition for resources. When extreme events induce stress, resources can become scarce in areas with high biomass and competition can suddenly drive mortality, overwhelming the benefits of cohabitation. Whether or not higher biodiversity protects an ecosystem from an extreme is highly site-specific.

Some plants respond to drought by reducing photosynthesis and transpiration immediately; others can maintain business as usual for much longer, stabilizing the response of the ecosystem as a whole. So the exact response of ecosystems to extremes depends on interactions between the type of event, plant strategies, vegetation composition and structure.

Which plant strategies will prevail is hard to predict and highly dependent on the duration and severity of the climatic extreme, and on previous extremes. Researchers cannot fully explain why some forests, tree species or individual plants survive in certain regions hit by extreme climate conditions, whereas entire stands disappear elsewhere. One study of beech trees in Germany showed that survival chances had a genomic basis, yet it is not clear whether the genetic variability present in forests will be sufficient to cope with future conditions.

And it can take years for ecosystem impacts to play out. The effects of the two consecutive hot drought years, 2018 and 2019, were an eye-opener for many of us. In Leipzig, tree growth declined, pathogens proliferated and ash and maple trees died. The double blow, interrupted by a mild winter, on top of the long-term loss of soil moisture, led to trees dying at 4-20 times the usual rate throughout Germany, depending on the species. The devastation peaked in 2020.

Ecosystem changes can also affect atmospheric conditions and climate. Notably, land-use change can alter the brightness (albedo) of the planet's surface and its capacity for heat exchange. But there are more-complex mechanisms of influence.

Vegetation can be a source or sink for atmospheric substances. A study published in 2020 showed that vegetation under stress is less capable of removing ozone than are unstressed plants, leading to higher levels of air pollution. Pollen and other biogenic particles emitted from certain plants can induce the freezing of supercooled cloud droplets, allowing ice in clouds to form at much warmer temperatures, with consequences for rainfall. Changes to species composition and stress can alter the dynamics of these particle emissions. Plant stress also modifies the emission of biogenic volatile organic gases, which can form secondary particles. Wildfires—enhanced by drought and monocultures—affect clouds, weather and climate through the emission of greenhouse gases and smoke particles. Satellite data show that afforestation can boost the formation of low-level, cooling cloud cover by enhancing the supply of water to the atmosphere.

Self-assessment

Read the self-assessment guidelines for *Self-assessment scale for reading comprehension* and *Self-assessment scale for written expression* (based on the *China's Standards of English Language Ability*) outlined in the appendix of this textbook. Please conduct a self-assessment to evaluate your own skills respectively.

Unit 3 Forest Ecosystem

Learning Objectives

After learning this unit, you will be able to:

- Understand and discuss the fundamental concepts and principles of ecological engineering as well as the role of forest in it.
- Make use of particular strategies and techniques to improve your efficiencies while reading for specific details.
- Identify and apply the skill of exemplification in writing exposition essays.

Part I Reading for Academic Purposes

Reading for Specific Details

Reading for details can be a daunting and challenging task but proves important in universities and other academic settings. Not only do we need to read critically, but we also need to read with an eye to distinguish fact from opinion and identify credible sources. Time is valuable, so reading things repeatedly is impractical. In universities, it is not unheard of to be assigned 500 pages of reading a week for just one of your classes. It literally may not be possible to read everything that is assigned, but you need to engage yourself in the reading and make a wise reader with proper strategies or techniques.

Reading for specific details, or detailed reading, is when you focus on the written material, really looking to gather specific information or evidence on a topic. This type of reading will provide you with a more indepth understanding of the specific information, facts, positions and views on a topic. In detailed reading you may be looking for new information or a different perspective. Also, try to consider all the possible perspectives of a subject as well as the potential for misunderstanding due to personal biases and the availability of false information about the topic.

Bear in mind what specific information you are looking for in the text is of vital importance in effective reading, especially for specific information.

Skimming and scanning are not only categorized as two of the ways of reading, they are also employed as reading strategies and skills in developing reading comprehension. The combination of both skills/strategies plays an essential role in achieving decent detailed reading ability.

Staying engaged in what you are reading is vital to reading for details. Annotating the writing as you read forces you to pay close attention and respond to what you are reading, consequently increasing your reading comprehension. Annotating can include noting anything specific that you are looking for (e.g., key words and phrases), anything that seems particularly significant, and any questions that you may have.

Finally, looking for particular patterns (words or expressions) in the writing can not only keep you engaged but also reveal more information about the writing. To do this, you need to build up a strong vocabulary and grammar to help you identify relevant or targeted information in the text.

Reading Passage 1

Critical Thinking Questions

Directions: Read the following passage and answer the following questions.

1. According to the author, how the study of ecological engineering is currently recognized in the academia?
2. How is the development of ecological theories different from that of experimental science?
3. What does Hubbert's "peak oil" curves reveal about the development of ecology?
4. What services else does nature offer to human being apart from food and clothing?

Concepts and Principles of Ecological Engineering

William J. Mitsch

Overview

A. Ecological engineering, defined as the design of sustainable ecosystems that integrate human society with its natural environment for the benefit of both, has developed over the last 30 years, and rapidly over the last 10 years. Its goals include the restoration of ecosystems that have been **substantially** disturbed by human activities and the development of new **sustainable** ecosystems that have both human and ecological values. It is especially needed as conventional energy sources **diminish** and amplification of nature's ecosystem services is needed even more. There are now several universities developing academic programs or departments called ecological engineering, ecological restoration, or similar terms, the number of **manuscripts** submitted to the journal *Ecological Engineering* continue to increase at a rapid rate, and the U.S. National Science Foundation now has a specific research focus area called ecological engineering. There are many private firms now developing and even **prospering** that are now specializing in the restoration of streams, rivers, lakes, forests, grasslands, and wetlands, the **rehabilitation** of minelands and urban brownfields, and the creation of treatment wetlands and phytoremediation sites. It appears that the perfect **synchronization** of academy, publishing research resources and practice is beginning to develop. Yet the field still does not have a formal **accreditation** in engineering and receives guarded acceptance in the university system and workplace alike.

B. *Principles of ecological engineering and related fields* have been published by Mitsch and Jorgensen, Mitsch, Straskraba, Zalewski, Bergen et al. and Odum. In fact, more principles have been developed for this field than many others. Yet there seems to be a continual **hue** and cry for yet more principles. Some of the more basic concepts, some of which could be principles as well, are self-design, acid-test, systems thinking, natural energy use, and ecosystem conservation.

1. Self-design

C. Self-design is one of the basic **cornerstones** of ecological engineering. It is the application of self-organization in the design of ecosystems. Nature contributes to the final design of ecosystems as much or more than does the human designer. This has been one of the most consistently used ecological engineering principles as "design" is one of the most important words in engineering.

2. Acid test of ecological theories

D. Creating or restoring an ecosystem is not usually experimental science. So few if any general scientific principles can be developed from ecological engineering. But when an ecosystem is created or restored, general principles already in existence in the field of ecology can be disproved. Bradshaw, who has described the restoration of a disturbed ecosystem as the "acid test of our understanding of that system" has stated that because we cannot prove that a restored ecosystem proves an ecological theory, we will "learn more from our failures than from our successes since a failure clearly reveals the inadequacies in an idea, while a success can only **corroborate** and support, and can never absolutely confirm, an **assertion**." Cairns was more direct in this point: "One of the most **compelling** reasons for the failure of theoretical ecologists to spend more time on restoration ecology is the **exposure** of serious weaknesses in many of the widely accepted theories and concepts of ecology."

3. System thinking

E. As described by Mitsch and Jorgensen in 2004, system thinking is required when ecosystems are created or restored. It is not the time to think about linear cause and effects but rather the ecosystem as a whole.

4. Natural energy use

F. Traditional engineering, by its very nature, depends on the energy in society. It has its successes partially due to the fact that seemingly any problem can be solved if enough energy is focused on solving that problem. "Engineering" and "energy" even have the same etymological roots (the origins or foundational elements of words). Ecological engineering uses self-designing ecosystems at its core so it is, by definition, a field that focuses on natural energies, often solar, wind, and **hydrologic** energies that have already been **manifest** in ecosystems. We now have a formal term—ecosystem services—for the human values that nature provides.

G. A fundamental concern in society is that we are now running out of energy to fuel our future. While there are of course claims to the contrary, it is now considered **conceivable** that we simply will not have as much energy resources in the future as we have had in the past. In a graphic similar to the all-too-accurate **projections** by Hubbert on the production and consumption of specific energy resources such as oil (the graphics became known as "peak oil" curves), Clugston, as one example, projected that it may be that we are going to go through a "peak energy" of human society of 570–710 Quads (quadrillion British Thermal Units (BTU)/year) by 2025–2030. The approaches and principles of ecological engineering need to be established now so that we can call on the ecosystem services of nature even more when we need them on the other side of the energy peak. As described by Day et al.

> We believe that in coming decades the restoration and sustainable management of rich natural ecosystems will be equally as important as the protection of existing wild areas. It will be a different kind of conservation because restored ecosystems will exist in a mosaic of intensively used areas, such as agroecosystems…

H. In a time of resource **scarcity**, especially energy, we suggest that ecological engineering (sometimes referred to as ecotechnology), including agroecology, is an appropriate basis for sustainable ecosystem management. Probably one of the most important shifts is for ecology to become more **prescriptive** and less **descriptive**, mostly through the growth of the ecological fields of ecological engineering and ecosystem restoration. Ecologists have a rich history of describing ecosystems and their functions but are less well trained in solving ecological problems. These new fields relate to solving ecological problems, borrowing approaches from engineering and landscape architecture. There are many active efforts in ecological engineering around the world …

5. Ecosystem conservation

I. Nature provides many valuable ecosystem services for humans. Of course, many of these values come from the harvest of plants and animals for food and fiber. The **identification** of these as well as many non-market values of nature, as for example illustrated by Costanza et al. has led to an increased emphasis on conservation by illustrating that nature has value. The development of new sustainable ecosystems and their values by ecological engineering will have the same effect.

J. Ecological engineers, then, have in their toolboxes all of the ecosystems, communities, populations, and organisms that the world has to offer. Therefore, a direct consequence of ecological engineering is that it would be **counterproductive** to eliminate or even disturb natural ecosystems unless absolutely necessary. This is **analogous** to the conservation ethic that is shared by many farmers even though they may **till** the landscape. In short, recognition of ecosystem values provides greater **justification** for the conservation of ecosystems and their species.

K. Aldo Leopold, the great Midwestern USA conservationist, stated this concept much more **eloquently**. As compiled by Aldo's son Luna Leopold after his father's death:

> The last word in ignorance is the man who says of an animal or plant: "What good is it?" If the land mechanism as a whole is good, then every part is good, whether we understand it or not. If the **biota**, in the course of **eons**, has built something we like but do not understand, then who but a fool would discard seemingly useless parts? To keep every cog and wheel is the first precaution of intelligent **tinkering**.

Note:

Peak oil refers to the hypothetical point at which global crude oil production will hit its maximum rate, after which production will start to decline. This concept is derived from geophysicist Marion King Hubbert's "peak theory", which states that oil production follows a bell-shaped curve.

Key Words and Phrases

1. substantially /səb'stænʃəli/	*adv.*	to a large degree 明显地；本质上
2. sustainable /sə'steɪnəb(ə)l/	*adj.*	able to continue over a period of time 可持续的
3. diminish /dɪ'mɪnɪʃ/	*v.*	to reduce or be reduced in size or importance 减少；降低
4. manuscript /'mænjuskrɪpt/	*n.*	the original copy of a book or article before it is printed 手稿；原稿
5. prosper /'prɒspə(r)/	*v.*	(of a person or a business) to be or become successful, especially financially 发达；兴旺
6. rehabilitation /ˌriːhəbɪlɪ'teɪʃ(ə)n/	*n.*	returning something to a good condition 恢复
7. synchronization /ˌsɪŋkrənaɪ'zeɪʃ(ə)n/	*n.*	the fact of happening at the same time, or the act of making things happen at the same time 同步发生

8. accreditation /əˌkredɪ'teɪʃ(ə)n/	*n.*	the fact of being officially recognized, accepted, or approved of 认可，认证
9. hue /hjuː/	*n.*	(a degree of lightness, darkness, strength, etc. of) a colour 色调；倾向
10. cornerstone /'kɔːnəstəʊn/	*n.*	something of great importance that everything else depends on 奠基石
11. corroborate /kə'rɒbəreɪt/	*v.*	to add proof to an account, statement, idea, etc. with new information 追加证明
12. assertion /ə'sɜːʃ(ə)n/	*n.*	a statement that you strongly believe is true 断言；论断
13. compelling /kəm'pelɪŋ/	*adj.*	if a reason, argument, etc. is compelling, it makes you believe it or accept it because it is so strong 令人信服的
14. exposure /ɪk'spəʊʒə(r)/	*n.*	the fact of something bad that someone has done being made public 公开；示众
15. hydrologic /ˌhaɪdrə'lɒdʒɪk/	*adj.*	of or related to the use of water on the earth 水文的
16. manifest /'mænɪfest/	*adj.*	easily noticed or obvious 明晰的，显然的
17. conceivable /kən'siːvəbl/	*adj.*	possible to imagine or to believe 可以理解和接受的
18. projection /prə'dʒekʃ(ə)n/	*n.*	a calculation or guess about the future based on information that you have 预测
19. scarcity /'skeəsəti/	*n.*	a situation in which something is not easy to find or get 缺乏；不足
20. prescriptive /prɪ'skrɪptɪv/	*adj.*	saying exactly what must happen, especially by making a rule and showing a certain degree of involvement 规范性的，规定的
21. descriptive /dɪ'skrɪptɪv/	*adj.*	saying what sb./sth. is like; describing sth. 描述性的，叙述

性的

22. identification /aɪˌdentɪfɪ'keɪʃ(ə)n/	*n.*	the ability to or act of recognize someone or something 认识；了解
23. counterproductive /ˌkaʊntəprə'dʌktɪv/	*adj.*	having an effect that is opposite to the one intended or wanted 产生反作用的
24. analogous /ə'næləgəs/	*adj.*	having similar features to another thing and therefore able to be compared with it 类似于
25. till /tɪl/	*v.*	to prepare and use land for growing crops 耕作土地
26. justification /ˌdʒʌstɪfɪ'keɪʃ(ə)n/	*n.*	a good reason or explanation for something 理由
27. eloquently /'eləkwəntli/	*adv.*	in a way that gives a strong, clear message 雄辩地
28. biota /baɪ'əʊtə/	*n.*	the animals and plants living in a particular place, time, or habitat 生物群
29. eon /'iːɒn/	*n.*	a period of time of one thousand million years or a period of time that is so long that it cannot be measured 千万年；永久
30. tinker /'tɪŋkə(r)/	*v.*	to make small changes to something, especially in an attempt to repair or improve it 修补；改善

Reading Comprehension

Directions: Read **Passage 1** again and answer the following questions.

1. In this passage, which of the following fields is not mentioned in recognizing the significance of ecological engineering?

A. Businesses. B. Government. C. Education. D. Publication.

2. Why is the concept of design included as one of the most important principles in ecological engineering?

A. Because self-design is one of the basic foundations of ecology.

B. Because nature makes more contribution to the formation of ecosystems than does the human.

C. Because of the fundamental components in the study of engineering.

D. Because ecological engineering is not classified as experimental science.

3. The principle of "natural energy use" indicates that ______.

A. almost all problems can be solved if enough energy is provided

B. human society is going through peak energy sooner or later

C. human are better to play a more active role in dealing with ecological engineering and ecological restoration

D. the restoration and sustainable management of natural ecosystems is anticipated to be as important as the protection of existing wild areas

4. The principle of "ecosystem conservation" denotes that ______.

A. humans sometimes have to sacrifice ecosystem in order to harvest plants and animals

B. farmers have not done a good job

C. nature has value because it provides human with food and clothing

D. natural ecosystems should be disturbed as less as possible

5. In Para. K, Luna Leopold's words "If the biota, in the course of eons, has built something we like but do not understand, then who but a fool would discard seemingly useless parts" refers to the concept that ______.

A. to keep the ecosystem conservation intact is the best policy for human

B. nature provides many valuable ecosystem services for humans, especially in the form of animals and plants

C. ecologists are expected to become more prescriptive and less descriptive, mostly through the development of ecological engineering and ecosystem restoration

D. people learn more from their failures than from their successes since a failure clearly reveals the inadequacies in the due idea

Vocabulary Exercises

Exercise 1

Directions: In this section, there are ten sentences, each with one word missing. You are required to complete these sentences with the proper form of the words given in the brackets.

1. Many theoretical ecologists are reluctant to spend more time on restoration ecology because serious weakness in many established understanding in ecology will be ____________ in the process. (exposure)

2. Ecological engineering increasingly meets the needs of current human society because traditional resources are ____________. (diminish)

3. True understanding of ecosystem values is crucial to ____________ the conservation of ecosystems. (justification)

4. Ecological engineering contains restoration of ecosystems which has been damaged due to ____________ human disturbance. (substance)

5. Failures work better than successes in developing ecological theories because the former surely point to what is ____________ in a hypothesis. (adequate)

6. When we humans fully ____________ the fact that nature provides us with multiple ecological services including both commercial and non-commercial values, we begin to lay more emphasis on ecological conservation. (identification)

7. The proportions of wheat ____________ by different livestock classes were assumed to be consistent over the years of the study. (consume)

8. The board chair said the highway was currently seeing between 20,000 and 24,000 vehicles a day, fewer than the 26,000 originally____________. (projection)

9. This piece of land has been ____________ for hundreds of years. (till)

10. A ____________ answer was provided in the final session from two different sources. (compel)

Exercise 2

Directions: In this section, there are ten sentences with ten blanks. You are required to select one word for each blank from a list of choices given in a word bank. Each choice in the bank is identified by a letter. You may not use any of the words in the bank more than once.

A. accreditation	B. amplification	C. analogous	D. assertion
E. conceivable	F. cornerstone	G. corroborate	H. manifest
I. descriptive	J. counterproductive	K. manuscript	L. prescriptive
M. prosper	N. scarcity	O. tinker	

1. Linguists have been confused by campaigns against sexist language: are they ____________ conspiracies or a kind of natural change?

2. The emergency vehicle for the International Space Station is ____________ to a lifeboat.

3. Not in all cases though: it is perfectly ____________ to think that donating a liver lobe is less burdensome whilst alive than once dead.

4. If the work is successful, spiritual well-being of connection is ____________ as appreciation for life, love of others, and feeling connected to deceased loved ones.

5. The sound ____________ seems to bring the listeners closer to the source.

6. As a way to improve traffic, widening roads can be ____________, as it may just encourage more people to drive.

7. Meaningful analysis of female occupation and illiteracy is thus non-viable because of the ____________ of data.

8. The hospital was threatened with the loss of ____________ if it did not improve the quality of its care.

9. All the parties to the dispute agree that ready access to the law is a ____________ of democracy.

10. Evolution cannot invent something quite new but can only ____________ with what is already there.

Exercise 3

Directions: Read the passage again and translate the following sentences from **Passage 1** (underlined in the passage) into Chinese.

1. Bradshaw, who has described the restoration of a disturbed ecosystem as the "acid test of our understanding of that system" has stated that because we cannot prove that a restored ecosystem proves an ecological theory, we will "learn more from our failures than from our successes since a failure clearly reveals the inadequacies in an idea, while a success can only corroborate and support, and can never absolutely confirm, an assertion".

2. One of the most compelling reasons for the failure of theoretical ecologists to spend more time on restoration ecology is the exposure of serious weaknesses in many of the widely accepted theories and concepts of ecology.

3. Probably one of the most important shifts is for ecology to become more prescriptive and less descriptive, mostly through the growth of the ecological fields of ecological engineering and ecosystem restoration.

Part II Academic Writing Strategies

Exemplification/Giving Examples

In academic writing, exemplification (or illustration as mentioned earlier in Chapter 1), which means "to give examples", is an essential and efficient strategy to explain or support your statements. It also adds interest to your writing.

To give strong support, you need to include examples that are only relevant to your statements. For instance, if you want to prove that taking public transportation can improve the environment, it is no help writing about the convenience you can have, such as reading while taking the train. Instead, you could give the data of the amount of emission decreased when people take train compared to driving cars.

In addition, to engage your audience, you want to vary the types of examples. Readers will easily become bored if you keep listing endless functions of forests in an attempt to illustrate the importance of forests. You can mention the process of how forests help to conserve water and soil, or even include some stories of how people mentally rely on forests. Here are some types of examples that can help with your writing: (1) people, place, actions and things; (2) facts and events; (3) stories or anecdotes.

Reading Passage 2

Critical Thinking Questions

Directions: Read the following passage and answer the following questions.

1. In what particular way(s) can forest serve human community?
2. What would you compare forest to as its role in maintaining earth's environment for the existence of human being?
3. What word(s) can be used to summarize the features of forest dwellers' indigenous knowledge?
4. What constitute the fundamental cause of forest degradation?

Forests: Ecological, Economic and Cultural Services

Mohammad Tareq Hasan

A. Forests and human beings are closely related with each other from the very beginning of the human history. Great social value of forests and their many ecological and economic services **render** significant contribution towards maintaining life conditions on earth. However, forest resources are being **depleted** at a great pace worldwide causing increased threat for living conditions on this planet. In the face of continuing deforestation (at 5.2 million hectares worldwide per year) the theme of World Environment Day 2011 "Forests: Nature at Your Service" appeared rather challenged. At the **juncture** of global climate change and rapid deforestation in Bangladesh (the annual rate in Bangladesh is 3.3 percent which is 0.6 percent in South Asia), it is now high time to look back at the ecological, economic and cultural roles played by forests. So that appropriate measures are taken before it is too late.

B. In Bangladesh 1.46 million hectares—17.5 percent of the country's land area—are under forest cover. However, canopy coverage would not be more than 6 percent. Forests of Bangladesh cover three major vegetation types, i.e. hill forests (evergreen to semi-evergreen); Sal forests (tropical moist deciduous) and Sundarbans (mangrove). Services provided by forests cover a wide array of ecological, economic, socio-cultural considerations and processes. However, these categories and groupings are neither **exhaustive** nor **discrete**.

Ecological services

C. The ecological services of forests are those environmental or ecological processes, which directly benefit humans. Some of the key ecological services are: carbon storage and **sequestration**, preservation and protection of **hydrological** function and conservation of biodiversity. Plants absorb carbon through photosynthesis from atmospheric carbon dioxide and return oxygen to the environment. Thus, simply being there forests reduce and keep carbon out of the atmosphere and maintain the earth's suitability for living. Therefore, forests can be deemed as the lungs of the earth.

D. Forests have major effects on hydrological processes also. Forests having large capacity for water absorption and **retention**, may sometimes convert irregular **precipitation** into a more even flow of water from **catchment** areas. The risk of flooding due to extreme weather and rainfall may, therefore, be reduced if forests exist there.

E. Moreover, forests are key components of biodiversity both in themselves and as a habitat for other species. Forests provide some of the most biodiversity rich ecosystem on earth and are supposed to provide habitat for an estimated 90 percent of the threatened and endangered species. Regardless of poor canopy coverage, forests of Bangladesh are rich in biodiversity. About 5,700 species of vascular plant including 300 tree species are found in Bangladeshi forests. There are approximately 840 wildlife species in the forests of Bangladesh, which includes 19 amphibian, 124 reptile, 578 bird and 119 mammal species. <u>Biodiversity has **intrinsic** value as well as provides practical and economic benefits forming the foundation of forest **dwelling** people.</u>

Economic services

F. Forests form the basis of a variety of industries including timber, processed wood and paper, rubber, fruits, etc. In Bangladesh 40% of the commercial timber is supplied by the Chittagong hill forests. Besides, different kinds of bamboo, undergrowth of the hill forests, is the most important raw material for paper **mills**. It is also used for house construction and supports many cottage industries. Sal forests of the central part of the country also provide economically valuable timbers.

G. In Bangladesh, Sundarbans—world's largest mangrove forest—provides livelihood directly to approximately 600,000 people who work as fishermen, wood-cutters (bawali), loggers, honey and wax collectors (mawal), etc. Forests also contain products that are necessary for communities living surrounding and depending on the forests. These products include: fuel, fodder, game, fruits, house building materials, medicines and herbs. Approximately 3 million people live in the villages surrounding the Sundarbans and are dependent solely on the forest resources for livelihood.

Socio-cultural services

H. Forests are home to millions of people world-wide, and many of these people are dependent on the forests for their survival. In addition, many people have strong cultural and spiritual **attachments** to the forests. The Munda, first settlers around the Sundarbans consider themselves part of the forest. They believe the forest to be the most holy place.

I. The issue of **indigenous** knowledge is also important. Many local people understand how to conserve and use forest resources in a sustainable way because of their continued attachment with forests over many years. It has often been argued that forests currently are being destroyed, in part because of the non-forest dwellers' lack of knowledge about the manner to best exploit the vast diversity of medicines, foods, natural fertilizers and pesticides that forests contain. For example: the wood-cutters and honey collectors of the Sundarbans have developed traditional cultural practices for **customary** use of resources. The traditional cultural practices of the golpata collectors do not permit to harvest an area more than once in a year. They only cut the leaves that are approximately nine feet long and they never destroy flowers and fruits. Honey collectors (mawal) make sure that young bees are never killed. Sheer dependence of these people on the forests has developed and fostered these norms and knowledge, which are an integral part of the forest-dependent peoples' cultures.

J. Non-forest dwellers are also increasingly exploiting forests now. It is now **invariably** agreed that forests keep the harmony **intact** between ecology, economy and culture. However, forests' sustainability seems to be **inversely** related to the population density. The main causes of degradation of forests and deforestation are over-exploitation due to population pressure, **encroachments**, shifting cultivation, **meager** employment opportunities outside agriculture, etc. So to **combat** over-**extraction** of forest resources and deforestation and protect the remains of natural forests and biodiversity people's participation in forest management must be increased. In this regard, expansion of agroforestry, training of local people and flow of information about the natural services provided by forests and how to best utilize the natural services in sustainable manners may play vital roles. Now is the time to act so that continued ecological, economic and social services are received from forests.

Key Words and Phrases

1. render /'rendə(r)/ *v.* to cause someone or something to be in a particular state 使成为
2. deplete /dɪ'pliːt/ *v.* to reduce something in size or amount, especially supplies of energy, etc. 耗尽；枯竭
3. juncture /'dʒʌŋktʃə(r)/ *n.* a particular point in time 时刻；关头
4. exhaustive /ɪg'zɔːstɪv/ *adj.* complete and including everything 彻底的
5. discrete /dɪ'skriːt/ *adj.* clearly separate or different in shape or form 明晰的

6. sequestration /ˌsiːkwes'treɪʃ(ən)/ *n.* the act of separating and storing a harmful substance such as carbon dioxide in a way that keeps it safe 隔离
7. hydrological /ˌhaɪdrə'lɒdʒɪkəl/ *adj.* relating to the study of water on the earth 水文学的
8. retention /rɪ'tenʃ(ə)n/ *n.* the continued use, existence, or possession of something 保持；持有
9. precipitation /prɪˌsɪpɪ'teɪʃ(ə)n/ *n.* water that falls from the clouds towards the ground, especially as rain or snow 降水
10. catchment /'kætʃmənt/ *n.* the area of land from which water flows into a river, lake, or reservoir 贮水区
11. intrinsic /ɪn'trɪnzɪk/ *adj.* being an extremely important and basic characteristic of a person or thing 内在的；固有的
12. dwell /dwel/ *v.* to live in a place or in a particular way 居住
13. mill /mɪl/ *n.* a factory where a particular substance is produced 工厂
14. attachment /ə'tætʃmənt/ *n.* a feeling of love or strong connection to someone or something 情感依恋
15. indigenous /ɪn'dɪdʒɪnəs/ *adj.* used to refer to, or relating to, the people who originally lived in a place, rather than people who moved there from somewhere else 原住民的
16. customary /'kʌstəməri/ *adj.* usual 习惯的
17. invariably /ɪn'veəriəbli/ *adv.* always 一致地；全然地
18. intact /ɪn'tækt/ *adj.* not damaged or destroyed 完好无损的
19. inversely /ɪn'vɜːsli/ *adv.* in the opposite way to something else 反向地
20. encroachment /ɪn'krəʊtʃmənt/ *n.* the act of gradually taking away someone else's rights or territory 侵占
21. meager /'miːgə(r)/ *adj.* (of amounts or numbers) very small or not enough 贫乏的；不足的
22. combat /kəm'bæt/ *v.* to try to stop something unpleasant or harmful from happening or increaseing; fight 与……斗争；解决；阻止
23. extraction /ɪk'strækʃ(ə)n/ *n.* the process of removing something, especially by force 开采；榨取

Exercise 1

Directions: Read the following sentences from **Passage 2** and rewrite the following sentences (underlined in the passage), replacing the underlined parts with your own words.

1. Great social value of forests and their many ecological and economic services render significant contribution towards maintaining life conditions on earth.

2. Biodiversity has intrinsic value as well as provides practical and economic benefits forming the foundation of forest dwelling people.

3. It is now invariably agreed that forests keep the harmony intact between ecology, economy and culture. However, forests' sustainability seems to be inversely related to the population density.

Exercise 2

Directions: Read **Passage 2** again and answer the questions below. Please give brief answers in about 10 words.

1. In what context was the theme of World Environment Day 2011 under pressure? (Para. A)

2. What's the percentage of Bangladeshi land area covered by forests? (Para. B)

3. What are the industries related to forests in Bangladesh? Please name at least four industries. (Para. F)

4. What population is solely dependent on the resources of Sundarbans? (Para. G)

5. What can be the solutions to combat over-extraction of forest resources? (Para. I and Para. J)

Exercise 3

Directions: Statements 1—5 (Table 5) are extracted from **Passage 2**. Please underline and identify the type of the example, if any, that support each of the statements.

Table 5 Examples of Statements

Statement	Details
Statement 1	Forests have major effects on hydrological processes also. (Para. D)
Statement 2	Moreover, forests are key components of biodiversity both in themselves and as a habitat for other species. (Para. E)
Statement 3	Forests are home to millions of people world-wide, and many of these people are dependent on the forests for their survival. (Para. H)
Statement 4	...many people have strong cultural and spiritual attachments to the forests. (Para. H)
Statement 5	It has often been argued that forests currently are being destroyed, in part because of the non-forest dwellers' lack of knowledge about the manner to best exploit the vast diversity of medicines, foods, natural fertilizers and pesticides that forests contain. (Para. I)

Exercise 4

Directions: Please write an example to support the statement(s) in **Exercise 3** that do(es) not have any example. Remember, to keep your writing engaging, you need to use different types of examples, and only relevant ones.

Extensive Reading and Writing

Directions: In this section, you will read five types of service from forests. Please employ a specific example from other geographical locations to illustrate one of the types. For instance, **Passage 2** describes the economic benefits of rainforests by taking the example of the Chittagong hill forests in Bangladesh, such as how this forest provides jobs and influences the local economy. You should write at least 120 words but no more than 180 words.

Five Types of Rainforest Ecosystem Services that Nourish People and Planet

Loulila Fenton

According to Conservation International's 2009 book, *The Wealth of Nature*, ecosystems support and regulate all natural processes on earth, while contributing to cultural, social, and economic benefits to human communities. These have become known as ecosystem services and, according to the Rainforest Conservation Fund (RCF), they would cost trillions of dollars per year if human beings had to provide them for themselves. Here are just five types of many of the ecosystem services provided to people and planet by the world's rainforests:

Supporting. The rainforest supports a number of natural cycles and processes. According to RCF, for example, many tropical rainforests live "on the edge"—they receive very few inputs of nutrients from the outside. This means that they have to produce most nutrients themselves. When a rainforest is whole it acts as a closed loop system and recycles the nutrients it has created. Without tree cover, these would be lost and the forest would not survive.

Soil formation is another important and related supporting service. Most rainforests are wet deserts—they are often formed in locations that normally cannot sustain much life. Trees and plants maintain soil quality by providing organic materials (leaves and branches). Their roots anchor the soil and thus prevent it, and the nutrients within it, from being washed away by high rainfall (soil erosion and nutrient leaching, respectively). This makes rainforest soils poorly adapted to agriculture since they are highly vulnerable to erosion and rapid loss of forest biodiversity—plethora of plant and animals found in the rainforest—once trees are removed.

Rain Making. Some forests' services extend across vast geographical areas. For example,

National Geographic reports that the Amazon rainforest makes as much as 50 percent of its own rainfall through a combination of processes. According to the Paradise Earth Project, Westward winds arriving in the Andes mountain range bring with them moisture from the Atlantic Ocean, which evaporates causing humidity. The plants use that water to grow and recycle it when they transpire—lose water into the atmosphere through their leaves. Each canopy tree transpires some 760 liters of water annually, translating to roughly 76,000 liters for every acre of canopy trees. This, and persistent cloud cover, add to humidity and form the basis of rains that move throughout the jungle. When these rains hit the rocky walls of the Andes, the water is deflected—a unique Amazonian characteristic that provides important rainfall that supports agriculture and water-based energy production in nearby parts of Brazil and Argentina. "The newly recognized Amazon rain machine is making a vital contribution to the Brazilian economy through its benefits to agro-industry and some hydro-electric facilities," says Paradise Earth.

Regulating. Rainforests help to maintain balance by regulating a number of processes. According to Rainforest Concern, "without rainforests continually recycling huge quantities of water, feeding the rivers, lakes and irrigation systems, droughts would become more common." Creation and recycling of rains thus helps to regulate the local climate.

Rainforests also help regulate air quality, whilst locking away carbon—in an opposite system to human beings, trees absorb (breathe in) atmospheric carbon dioxide and produce (breathe out) oxygen; they use the extracted carbon as raw material for growth of their living parts (stems, leaves, and roots).

Provisioning. Rainforests do not simply play supporting and regulating roles. They are also prolific producers of goods that people derive economic value from—according to the *United Nations Environment Programme (UNEP) 2011 Year Book*, about 1.6 billion people rely in some way on forests for their livelihoods. For example, when trees are extracted from them, rainforests provide human necessities such as wood, fiber, and fuel. In addition, species and processes of the rainforest provide an invaluable source of ideas for the growing field of biomimicry—the examination and emulation of nature to find solutions to human problems.

Beyond privately gained extractive benefits, rainforests are also vital for local, often indigenous, communities that inhabit them. According to *The Wealth of Nature* these include fresh water, wild foods, crops and livestock, wild fisheries, wood for fire and construction, fibers and other materials for arts and crafts, and natural biomedicines and pharmaceuticals.

Culture Sustaining. Rainforests are also crucial to culture and society. They are increaseingly popular destinations for recreation and eco-tourism. For those far away, they hold much educational and scientific value. For those living close, they are a source of a deep sense of belonging, cultural heritage, and religious and spiritual significance. And, of course, their beauty provides immeasurable aesthetic value to the world at large.

According to the *2011 Keeping Track of Our Changing Environment: From Rio to Rio+20 (1992—2012)* report by the UNEP, since 1990, the world's primary forest area has decreased by 300 million hectares, an area larger than Argentina. Yet, without the ecosystem services that forests provide, many natural and human processes would collapse. This scientific fact makes planet Earth's natural treasures worth saving.

Self-assessment

Read the self-assessment guidelines for *Self-assessment scale for reading comprehension* and *Self-assessment scale for written expression* (based on the *China's Standards of English Language Ability*) outlined in the appendix of this textbook. Please conduct a self-assessment to evaluate your own skills respectively.

Unit 4 Wetland Ecosystem

Learning Objectives

After learning this unit, you will be able to:

- Know about the importance of wetlands in the sustainability of environment.
- Acquire the development techniques of paragraphs.
- Learn the paragraph development method: comparison and contrast.

Part I Reading for Academic Purposes

Identifying Features of Paragraphs

All essays or articles are made up of paragraphs; paragraphs are, in turn, made up of sentences. Although paragraphs are made up of sentences, single, correct, and effective sentences put together do not necessarily make an effective paragraph. A good paragraph has one, and only one, central idea. A paragraph is a series of related sentences that support the central idea.

Effective organization of ideas in a paragraph is essential for clear communication. One key feature of a paragraph is UNITY. Probably the most effective way to achieve paragraph unity is to express the central idea of the paragraph in a topic sentence, which is the main point of the paragraph. A topic sentence has a unifying function, but a topic sentence alone doesn't guarantee unity. A paragraph is unified if all the sentences relate to the topic sentence. We should be clear that not all paragraphs need topic sentences. In particular, opening and closing paragraphs, which serve different functions from body paragraphs, generally don't have topic sentences. In academic writing, the topic sentence nearly always works best at the beginning of a paragraph so that the reader knows what to expect.

In many cases, a body paragraph demonstrates and develops the topic statement through an orderly, logical progression of ideas, which draws forth the second feature of an effective paragraph—COHERENCE. Paragraph coherence means the practice to bind sentences together so that all the ideas or thoughts in the paragraph can fit together in a logical and smooth flow, establishing a connection between the main sentence and all the other sentences. In this light, each body paragraph should better have a topic sentence, which provides information on the main focus and shows direction of the paragraph development. All the remaining sentences should support or develop the idea in the topic sentence. The last sentence in the paragraph, or the concluding sentence, may summarize the paragraph, or provide a link with the following paragraph.

Thus, we can say that an effective paragraph generally has three fundamental parts: a central idea, supporting evidence, and concluding sentence. When we write a paragraph, the first thing we should do is to think of a central idea or theme and write it down in a complete sentence, which will serve as the topic sentence. Then we should think of the details or examples to support the main idea. We need to work out an outline to put these details in good order. With the outline, the paragraph is almost half-done.

There are a number of useful techniques for expanding on topic sentences and developing ideas in paragraphs, such as illustration, definition, analysis or classification, comparison or contrast, cause and effect, and so on and so forth.

Reading Passage 1

Critical Thinking Questions

Directions: Read the following passage and answer the following questions.

1. What does the first paragraph show about urban wetlands?
2. What's the possible prospect of the dirty wetland if proper management is under way?
3. What should urban managers do to guarantee the biggest benefits provided by wetlands?

Urban Wetlands: The Challenge of Making Hidden Values Visible

Rob Mc. Innes

A. Across towns and cities in the developing world, the scene is all too common: a heavily polluted black ditch, **discarded** plastic bottles bobbing and amassing on the surface of the **contaminated** watercourse, untreated effluent and highway drainage seeping **ceaselessly** into the channel and the **stench** of methane and faeces **wafting** through the warm urban breeze. The backs of informal settlements fringe these polluted shores as urban populations **shun** the **unsightly** drains. As chronic pollution and environmental degradation hasten the **prevalence** of pest species, increase the potential for the spread of disease and undermine human health, often the prevailing wisdom results in infilling and the conversion of open channels to buried concrete pipes. Against this background, it is a paradox that these degraded urban drains, streams and rivers are often great places to observe local wetland birds such as night herons and egrets as they opportunistically

gorge themselves on the available smorgasbord of invertebrates, amphibians and oxygen starved fish.

B. This picture highlights one, if not the main, challenge facing urban wetlands. Often wetlands, however degraded or polluted, can continue to provide vital but unacknowledged functions and services which **underpin** human well-being. Yet, in many towns and cities unless there is an obvious and clearly recognized value, wetlands fail to be adequately considered in traditional decision-making and their further degradation and loss prevails. Through appropriate management interventions, the black ditch could be transformed into a multi-functional solution provider acting as a waste water treatment facility, a flood **retention** area, a natural air conditioner, a carbon store, a source of fuel and medicinal plants and also a wildlife habitat for a broad variety of species which could control pests and disease rather than a buried pipe.

C. The potential of wetlands to be considered as systemic solutions, or low-input technologies which use natural processes to optimize benefits across the spectrum of ecosystem services and their **beneficiaries** is still poorly integrated into urban management and planning. A review conducted on behalf of the United Nations Human Settlements Program (UN-Habitat) demonstrated that even the most biodiversity aware and green cities fail to recognize the full range of ecosystem services provided by natural systems within urban areas. This failure of urban managers and planners to recognize ecosystem services and the benefits they bring extends across different socioeconomic and geographical environments and is not just an issue in the developing world. The inability to appreciate the benefits provided by natural infrastructure is greatest for wetlands, where some of the essential services, such as the storage, recycling and processing of nutrients, the accumulation of organic matter, carbon storage and the role of pollination, are rarely visible in decision-making.

D. If key governmental **stakeholders** are not aware of the importance or prevalence of the vital services that wetlands perform, then their **preclusion** from and integration into decision-making is almost inevitable. However, the local knowledge and understanding of communities who directly interact with urban wetlands can often paint a different picture. Some immigrant workers in Chinese Taipei are often seen to be exploiting the free food available from the fish populations in wetlands constructed as water treatment and recreational areas along the banks of certain rivers. Similarly, in the wetlands of urban Colombo, Sri Lanka, local residents often collect medicinal plants, such as nika nika (Sesbania), for their own personal use and administration. However, there is a disconnect between the local knowledge and use and the information used by local governments in decision-making. Urban managers and planners need to improve a range of techniques, approaches or tools to ensure that the full range of benefits provided by wetlands is considered and incorporated into decision-making. A starting point for this is the simple recognition of the ecosystem services being provided. This can take the form of a simple **binary** (presence-absence) assessment of a list of potential ecosystem services and does not necessarily need to involve costly and time-consuming **monetization** exercises (which are often flawed as they fail to recognize the full range of values, use valuation techniques which utilize inappropriate value transfer methods, do not adequately take into account changing economic drivers over time and simplify complex ecological relationships towards reductive services with a market value). The implementation of a simple approach to recognizing ecosystem services, which incorporates local and city-wide values, is fully **commensurate** with the tiered basis of seeking to recognize, demonstrate and capture value of services as advocated through the work of The Economic of Ecosystems and Biodiversity and would raise the visibility of the multiple values of urban wetlands.

E. However, there are other factors that require consideration when trying to raise the visibility of the benefits that wetlands provide in urban areas. Often the promotion of systemic solutions such as multi-functional wetlands or the wider potential of blue-green infrastructure, is often perceived as "too good to be true" and hence automatically mistrusted. This attitude is often underpinned by a blind faith in the confidence behind traditional engineered solutions (such as building flood embankments or implementing electro-mechanical water treatment plants) to solve problems. Institutional structures and silos have been constructed in local governments across the world, and within the large **consultancies** upon which they depend for implementing projects, that prevent the cross fertilization of natural and man-made solutions. Consequently biodiversity in towns and cities has become the responsibility of nature conservation or environmental agencies which often focus on the rare, threatened and mysterious species and habitats and not on the **utilitarian** values of the lowest **trophic** levels, such as nematodes, earthworms, fungi and bacteria. The choice is often "either-or" not integration of both natural and engineered solutions. This is despite that fact that where the delivery of multiple ecosystem services has been the objective within an urban area it has been demonstrated that the delivery of **collateral** ecosystem services will flow from the protection, restoration and management of biodiverse habitats.

F. Therefore, it is worth remembering that cities are human constructs and, as such, humanity has the power to steer and alter their developmental **trajectories** if decision-makers are provided with the appropriate knowledge, understanding and capacity to alter existing management approaches. The failure to recognize the basic ecological relationships and value of natural infrastructure that support all ecosystem services raises serious concerns regarding the integration of the multiple benefits of natural systems into decision-making.

G. At the Eleventh Meeting of the Conference of the Parties to the Convention on Wetlands in 2012, Resolution XI.11 on the Principles for the planning and management of urban and peri-urban wetlands was adopted. These principles recognized that urban and peri-urban wetlands provide a range of ecosystem services, including providing food, water and fuel, improving water quality, assisting in water security, **mitigating** natural hazards through the regulation of flooding and reduction of storm surges, regulating local climate, as well as the positive contribution that wetlands can make to people's physical and mental well-being. They also acknowledged the importance of integrating local knowledge into decision-making and the need for governments to act upon these principles, further **disseminate** them and ensure that they are taken up by the sectors and levels of government responsible for the planning and management of urban and peri-urban environments.

Key Words and Phrases

1. discard /dɪ'skɑːd/	*v.*	to throw sth. out or away, to get rid of sth. that is no longer useful 扔掉，丢弃
2. contaminate /kən'tæmɪneɪt/	*v.*	to pollute, make sth. dirty, impure or harmful 污染，弄脏
3. ceaselessly /'siːsləsli/	*adv.*	without end 不停地
4. stench /stentʃ/	*n.*	a strong, very unpleasant smell 臭气；恶臭
5. waft /wɒft/	*v.*	(cause sth. to) be carried lightly and smoothly through the air（使某物）在空中漂浮或飘荡
6. shun /ʃʌn/	*v.*	to avoid sb./sth.避开；回避
7. unsightly /ʌn'saɪtli/	*adj.*	not pleasant to look at 难看的；不雅观的
8. prevalence /'prevələns/	*n.*	the state of being widespread 普遍；盛行；流行
9. gorge /gɔːdʒ/	*v.*	to eat lots of sth. in a very greedy way 狼吞虎咽
10. underpin /ˌʌndə'pɪn/	*v.*	to support or form the basis of 加强，巩固；构成（……的基础等）
11. retention /rɪ'tenʃ(ə)n/	*n.*	the action of keeping sth. rather than losing it 保持；维持；保留
12. beneficiary /ˌbenɪ'fɪʃəri/	*n.*	a person who gains as a result of sth. 受益者；受惠人
13. stakeholder /'steɪkhəʊldə(r)/	*n.*	a person or company that is involved in a particular organization, project, etc., especially because they have invested money in it（某组织、工程等的）参与人，参与方
14. preclusion /prɪ'kluːʒ(ə)n/	*n.*	the act of preventing sth. by anticipating and disposing of it effectively 排除；阻止；妨碍；预防

15.	binary /'baɪnərɪ/	*adj.*	of or involving a pair or pairs 成双的；包含两部分的；二元的
16.	monetization /ˌmʌnɪtaɪ'zeɪʃ(ə)n/	*n.*	establishing sth. (e.g. gold or silver) as the legal tender of a country 货币化
17.	commensurate /kə'menʃərət/	*adj.*	matching sth. in size, importance, quality, etc.（在大小、重要性、质量等方面）相称的，相当的
18.	consultancy /kən'sʌltənsi/	*n.*	a company that gives expert advice on a particular subject to other companies or organizations 咨询公司
19.	utilitarian /ˌju:tɪlɪ'teəriən/	*adj.*	designed to be useful rather than luxurious or decorative; severely practical 实用的（非奢华的或装饰的）；功利的
20.	trophic /'trəʊfɪk/	*adj.*	of or relating to nutrition 营养的，有关营养的
21.	collateral /kə'lætərəl/	*adj.*	accompanying as secondary or subordinate 附属的；次要的
22.	trajectory /'trædʒiktəri/	*n.*	the path followed by an object moving through space 轨道，轨线
23.	mitigate /'mɪtɪgeɪt/	*v.*	to make sth. less harmful, serious, etc. 减轻；缓和
24.	disseminate /di'semineit/	*v.*	to cause to become widely known 宣传，传播；散布

Reading Comprehension

Directions: Read **Passage 1** again and answer the following questions.

1. What is the seemingly contradictory phenomenon existing in urban areas in developing countries?

A. The severely polluted rivers flow endlessly into channels.

B. The wetlands birds feed on rubbish.

C. The dirty wetlands are also breeding spots for wild birds.

D. The worsening environment has resulted in large numbers of pests.

2. What does a review carried out on behalf of the United Nations Human Settlements Program demonstrate?

A. Only the most biodiversity-aware cities recognize the full range of ecosystem services.

B. There is a long way to go when it comes to the full range of ecosystem services.

C. The most green cities have succeeded in understanding the full range of ecosystem services.

D. The most biodiversity-aware and green cities have worked together to realize the full range of ecosystem services.

3. How did some local residents in the developing world make use of wetland animal and plant species?

A. They cared for the large numbers of animals and plants to protect environment.

B. They made use of the free food available from fish species.

C. They took advantage of some wetland animals and plants to benefit themselves.

D. They collected medicinal plants in their own interest.

4. What would happen if the authority couldn't realize the basic ecological relationships and value of natural infrastructure?

A. It would cause serious worries.

B. The benefits of natural systems could integrate into decision-making.

C. The delivery of collateral ecosystem services would flow from the protection.

D. People would have blind faith in the confidence behind traditional solutions.

5. Which of the following is not recognized as ecosystem services provided by wetlands at the Eleventh Meeting of the Conference of the Parties to the Convention on Wetlands?

A. Suppying food, water and fuel.

B. Contributing to water security.

C. Regulating local weather conditions.

D. Improving people's housing conditions.

Vocabulary Exercises

Exercise 1

Directions: In this section, there are ten sentences, each one with one word missing. You are required to complete these sentences with the proper form of the words given in the brackets.

1. More than half of those cases can be attributed to contact with ____________ water and a lack of proper sanitation. (contaminate)

2. The ____________ of several silkworm diseases has led to a decline in silk products. Especially in Europe, the silk industry has never recovered. (prevail)

3. He has been complaining about this or that ____________ all the time, which really gets on my nerves. (cease)

4. The reform of the political system has once encountered ____________ from riots and terrorists in some countries. (preclude)

5. Nothing is more ____________ than watching someone talk and chew their food at the same time. (sightly)

6. On the other hand, there are other factors that require consideration when trying to raise the ____________ of the benefits that wetlands provide in urban areas. (visible)

7. Increasing numbers of universities have been ____________ their courses by offering more options and adding communicative features to the current system. (optimum)

8. In many towns and cities unless there is an obvious and clearly recognized value, wetlands fail to be ____________ considered in traditional decision-making. (adequate)

9. Chinese economists have made enormous ____________ to the field of financial and corporate economics. (contribute)

10. If any of the updates requires you to restart the computer to complete the installation, Windows restarts the computer ____________. (automatic)

Exercise 2

Directions: In this section, there are ten sentences with ten blanks. You are required to select one word for each blank from a list of choices given in a word bank. Each choice in the bank is identified by a letter. You may not use any of the words in the bank more than once.

A. mitigating	B. exploiting	C. utilitarian	D. undermine
E. sustaining	F. infrastructure	G. accessible	H. opportunistically
I. foundation	J. classify	K. retention	L. integrated
M. independent	N. shun	O. discard	

1. Some western intelligence agencies are charged with trying to ____________ the government by providing information for rebels.

2. It is a paradox that these degraded urban drains, streams and rivers are often great places to observe local wetland birds as they ____________ gorge themselves on the available smorgasbord.

3. The potential of wetlands to be considered as systemic solutions, or low-input technologies which use natural processes to optimize benefits across the spectrum of ecosystem services and their beneficiaries is still poorly ____________ into urban management and planning.

4. These principles recognized that urban and peri-urban wetlands provide a range of ecosystem services, including providing food, water and fuel, improving water quality, assisting in water security, ____________ natural hazards, and so on.

5. Once we start to see people as individuals, and ____________ the stereotypes, we can move positively toward inclusiveness for everyone.

6. Some people are using bikes for a ____________ purpose or as a substitute for public transport when going on a short ship.

7. We should keep in mind that ____________ of knowledge requires concentration of the mind, so make sure that you won't get absent-minded when pursuing knowledge.

8. You can hardly ____________ meeting her if you and she both work in the same office despite the fact that you take great pains not to do so.

9. Some immigrant workers are often seen to be ___________ the free food available from the fish populations in wetlands constructed as water treatment and recreational areas along the banks of certain rivers.

10. The government will accelerate restoration and reconstruction of ___________, including transportation, communications, energy, water conservancy and distribution.

Exercise 3

Directions: Read the passage again and translate the following sentences from **Passage 1** (underlined in the passage) into Chinese.

1. If key governmental stakeholders are not aware of the importance or prevalence of the vital services that wetlands perform, then their preclusion from and integration into decision-making is almost inevitable.

2. Often the promotion of systemic solutions such as multi-functional wetlands or the wider potential of blue-green infrastructure, is often perceived as "too good to be true" and hence automatically mistrusted.

3. It is worth remembering that cities are human constructs and, as such, humanity has the power to steer and alter their developmental trajectories if decision-makers are provided with the appropriate knowledge, understanding and capacity to alter existing management approaches.

Part II Academic Writing Strategies

Comparison and Contrast in Paragraph Writing

Comparison and contrast is one of the frequently used methods to develop a paragraph. Comparison is the process of examining two or more things in order to establish their similarities, while contrast is the process of examining two or more things in order to unveil their differences. We might compare or contrast two products of different brands, two television shows, two instructors, two jobs, two cities, or two causes of action.

There are many directions a compare-and-contrast paragraph can take, but it should always make an argument that explains why it's useful to put these two subjects together in the first place. A good comparison and contrast paragraph will help your readers understand why it's useful or interesting to put these two subjects together: (1) to show that A is better than B. (e.g. traveling by car as compared with traveling by train), (2) to show differences between two seemingly similar items (e.g. yourself compared with your sister), (3) to show similarities between two seemingly different items (e.g. the education of a university as compared with the production of a machine), etc.

There are several ways to organize a comparison-and-contrast paragraph. Which one to choose depends on what works best for your ideas. The following elaborates on the two main formats.

(1) Subject to subject Pattern. This organization method is to write down all the points about one of the subjects to be compared/contrasted, and then to talk about all the main points about the other subject.

Introduction (includes topic sentence and sets up comparison/contrast)

Subject 1

Point 1

Point 2

Point 3

Subject 2

Point 1

Point 2

Point 3

Conclusion (summary and corresponding to the beginning)

Sample:

It is easy to be a winner. A winner can show his joy publicly. He can laugh and sing and dance and celebrate his victory. People love to be with winners. Winners are never lonely. But unlike winners, losers are the lonely ones of the world. It is difficult to face defeat with dignity. Losers cannot show their disappointment publicly. They cannot cry or grieve about their defeat. They may suffer privately, but they must be composed in public. They have nothing to celebrate and no one to share their sadness.

Comment:

In the above paragraph, the writer contrasts winners with losers. He first discusses winners, then losers. "But unlike winners" functions as a transition between the two and emphasizes the contrast.

(2) Point-by-point Pattern. This organization method switches back and forth between subject points. That is, the way we can write a paragraph that involves comparison/contrast is to take each point in turn, and contrast them immediately, as follows:

Introduction (includes topic sentence and sets up comparison/contrast)

Subject 1 vs Subject 2 Point 1

Subject 1 vs Subject 2 Point 2

Subject 1 vs Subject 2 Point 3

Conclusion (summary and corresponding to the beginning)

Sample:

Nation

Franklin D. Roosevelt

A nation, like a person, has a body—a body that must be fed and clothed and housed, invigorated and rested, in a manner that measures up to the standards of our time. A nation, like a person, has a mind—a mind that must be kept informed and alert, that must know itself, that understands the hopes and the needs of its neighbors—all the other nations that live within the narrowing circle of the world. And a nation, like a person, has something deeper, something more permanent, something larger than the sum of all its parts. It is that something which matters most to its future—which calls forth the most sacred guarding of its present.

Comment:

This extract is a typical comparison. A nation and a person are quite different, but they are comparable somehow. They are similar in three aspects: the body, the mind and something deeper—the spirit. The writer elaborates on the three aspects point by point very clearly.

Subject-to-subject comparison is often adopted when only a few aspects are discussed. Transition words like "like, unlike, as, compared with, in contrast to" are often used. These words remind readers of the subjects under discussion. If we want to mention a number of aspects, it is better to use the point-by-point pattern. In this pattern, we put all the similarities between the two subjects together and all the differences together. It's suited to subjects whose similarities and differences are more balanced. In short, no matter which method we use, the fittest is the best.

Reading Passage 2

Critical Thinking Questions

Directions: Read the following passage and answer the following questions.

1. Why is wise use of wetlands so vital to global sustainability?
2. How can the Convention on Wetlands realize the ambitious goals for wetlands?
3. How does the writer organize the passage?

Wise Use of Wetlands Is Critical to Global Sustainability

Marianne Courouble, Nick Davidson, Lars Dinesen, et al.

A. The 2030 Agenda for **Sustainable** Development, adopted in 2015, provides a blueprint for peace and prosperity for people and the planet. Conservation and wise use of wetlands is an important pathway to meeting many of the 17 goals and 169 targets of the SDGs (Sustainable Development Goals), with benefits also for global targets related to climate change, biodiversity conservation and disaster risk reduction. The Convention on Wetlands is, together with the UN Environment Programme, are co-custodian of SDG Indicator 6.6.1, and data submitted by Contracting Parties based on *National Wetland **Inventories*** is used to track change in extent of water-related ecosystems over time.

B. The **linkages** between wetlands and sustainable development outcomes are expressed in several ways, and actions for wetlands have an implication for delivery on SDGs much beyond SDG 6 Target 6. For example, maintaining healthy inland wetlands contribute to protecting the coastal environment from **eutrophication**, thereby contributing to delivering on SDG 14 Life Below Water, and also helps sustain fishery productivity, contributing to SDG 2 Zero Hunger. In 2017, fish consumption accounted for 17% of global population's intake of animal proteins and at least two-thirds of all the fish consumed worldwide are dependent on coastal wetlands.

C. Having access to quality blue environments, such as wetlands, can significantly benefit human health. Feeling psychologically connected to, living near, or undertaking recreation in the natural world such as wetlands is associated with better mental health.

D. Water and wetlands are "climate connectors" requiring **collaboration** and **coordination** across actions needed for sustainable development, climate change and disaster risk reduction, while wetlands are also critically important carbon stores. Healthy wetlands and **equitable** sharing of benefits can also contribute to peace making, though wetland conservation still does not figure prominently in peace-building efforts.

E. Nature-based solutions for water, which **incorporate** the role that healthy and sustainably managed wetlands play in buffering water-related risks, are an important pathway for moving beyond a "business as usual" focus on human-built infrastructure. This is particularly true for coastal areas, which carry a **disproportionately** high concentration of population and economic assets, higher rates of population growth, sediment deprived deltas and human-induced land **subsidence**. These together lead to the phenomenon of sinking, making the impacts of sea-level rise even more extreme. Though the uptake of nature-based solutions has increased recently, critical challenges remain in terms of rising investment and knowledge.

F. Considerable progress has been made towards targets **designating** portions of the planet's land and oceans as protected areas. However, further work is needed to ensure that these are ecologically representative and safeguard the most important areas for biodiversity. Protected and conserved areas need to be connected to one another as well as to wider land and seascapes and to be equitably and effectively managed. Existing **multilateral** environmental agreements provide a platform of **unprecedented** scope and ambition for action, but greater national commitment and effective cooperation in using and implementing these established **mechanisms** are vital to enable such international instruments to effectively safeguard ecosystems.

G. Finally, wetlands play critical, spiritual, **aesthetic** and cultural roles. Ecosystem services **encompass** far more than strictly utilitarian purposes. For example, thousands of **pilgrims** each year brave harsh weather to visit the high altitude, Himalayan wetland Mansarovar for spiritual **atonement**, one of **innumerable** sacred lakes, wells, springs and rivers. An important dimension of justice, for humans and the planet, is the recognition of "rights of nature" within legal frameworks, including proposals for a universal "Rights of Wetlands" statement. This puts the human species in a more respectful relationship with nonhuman nature for effective, sustainable and ethical "**stewardship** of the Earth and the life on it".

H. The changes needed to **stabilize** the environment over the next few years are profound and reach far beyond **conventional** ideas of conservation. Current development trajectories are insufficient to conserve and sustainably use nature and ensure the SDGs. Key **leverage** points for transformation towards sustainability identified in the IPBES process include directing efforts towards: (1) visions of a good life; (2) total consumption and waste; (3) values and action; (4) inequalities; (5) justice and inclusion in conservation; (6) **externalities** and linkages at a distance (so called "telecouplings"); (7) technology, innovation and investment; and (8) education and knowledge generation and sharing.

I. Wetland wise use supports the delivery of these leverage points and, when placed at the centre of decision making, helps to ensure sustainable development.

J. To reach these ambitious goals in wetlands, the Convention on Wetlands needs better implementation and more effective leveraging of the synergies that exist with other Multilateral Environmental Agreements (MEAs) and institutions. Wetlands of International Importance themselves are important not just as a means of increasing the chances of wise use in critically important wetlands, but also as laboratories for sustainable wetland management. Working with Contracting Parties, including capacity building to support national implementation, is a critical element of success.

Key Words and Phrases

1. sustainable /sə'steɪnəb(ə)l/	*adj.*	sth. that is sustainable is sth. that can continue or be continued for a long time 可持续的
2. inventory /'ɪnvəntri/	*n.*	a written list of all the objects or things to be done 详细目录；清单
3. linkage /'lɪŋkɪdʒ/	*n.*	action or manner of connecting things or being connected 联系；连接；结合
4. eutrophication /ˌjuːtrəfɪ'keɪʃ(ə)n/	*n.*	the process of too many plants growing on the surface of a river, lake, etc., often because chemicals that are used to help crops grow have been carried there by rain（由雨水带的化肥等造成水体的）富营养化
5. collaboration /kəˌlæbə'reɪʃ(ə)n/	*n.*	the act of working with another person or group of people to create or produce sth.合作；协作
6. coordination /kəʊˌɔːdɪ'neɪʃ(ə)n/	*n.*	the act of making parts of sth.,

		groups of people, etc. work together in an efficient and organized way 协作；协调；配合
7．equitable /'ekwɪtəb(ə)l/	*adj.*	fair and reasonable; treating everyone in an equal way 公平合理的；公正的
8．incorporate /ɪn'kɔːpəreɪt/	*v.*	to include sth. so that it forms a part of sth. 将……包括在内；吸收；使并入
9．disproportionately /ˌdɪsprə'pɔːʃənətli/	*adv.*	in a way that is out of proper relation with sth. else in size, number, importance, etc. 不相称地；不成比例地
10．subsidence /səb'saɪdns; 'sʌbsɪdns/	*n.*	the process by which an area of land sinks to a lower level than normal 下沉；沉降；下陷
11．designate /'dezɪgneɪt/	*v.*	to show sth. using a particular mark or sign 标明；标示；指明
12．multilateral /ˌmʌlti'lætərəl/	*adj.*	agreed upon or participated in by different parties, especially government of different countries 多边的；多国的
13．unprecedented /ʌn'presɪdentɪd/	*adj.*	never having happened, been done or been known before 无前例的；前所未有的；空前的
14．mechanism /'mekənɪzəm/	*n.*	a method or a system for achieving sth. 方法；机制
15．aesthetic /iːs'θetɪk/	*adj.*	concerned with beauty and art and the understanding of beautiful things 审美的；有审美观点的；美学的
16．encompass /ɪn'kʌmpəs/	*v.*	to include a large number or range of things 包含，包括，涉及
17．pilgrim /'pɪlgrɪm/	*n.*	a person who travels to a holy place for religious reasons 朝觐者；朝圣的人
18．atonement /ə'təʊnmənt/	*n.*	act of compensating for a previous wrong, error, etc.补偿；赎罪
19．innumerable /ɪ'njuːmərəb(ə)l/	*adj.*	too many to be counted 数不清的；无数的

20.	stewardship /'stju:ədʃɪp/	*n.*	the act of taking care of or managing sth.管理；看管；组织工作
21.	stabilize /'steɪbəlaɪz/	*v.*	to become or to make sth. become firm, steady and unlikely to change; to make sth. stable（使）稳定；稳固
22.	conventional /kən'venʃ(ə)n(ə)l/	*adj.*	following what is traditional or the way sth. has been done for a long time 传统的；习惯的
23.	leverage /'li:vərɪdʒ/	*n.*	the ability to influence what people do 影响力
24.	externality /ˌekstɜ:'næləti/	*n.*	the quality or state of being outside or directed toward or relating to the outside or exterior 外在性；外部事物

Writing Tasks

Exercise 1

Directions: Read the following sentences from **Passage 2** and rewrite the following sentences (underlined in the passage), replacing the underlined parts with your own words.

1. Water and wetlands are "climate connectors" requiring <u>collaboration</u> and <u>coordination</u> across actions needed for sustainable development, climate change and disaster risk reduction.

2. This is particularly true for coastal areas, which carry a <u>disproportionately</u> high <u>concentration</u> of population and economic assets, higher rates of population growth, sediment <u>deprived</u> deltas and <u>human-induced</u> land subsidence.

3. <u>Considerable</u> progress has been made towards targets <u>designating</u> portions of the planet's land and oceans as protected areas.

4. Thousands of <u>pilgrims</u> each year <u>brave</u> harsh weather to visit the high altitude, Himalayan wetland Mansarovar for spiritual <u>atonement</u>, one of <u>innumerable</u> sacred lakes, wells, springs and rivers.

Exercise 2

Directions: Read **Passage 2** again and answer the questions below. Please give brief answers in about 10 words.

1. What is an important factor in realizing many of the goals and targets of the SDGs? (Para. A)

2. How much fish eaten all over the world is dependent on coastal wetlands? (Para. B)

3. What is one of the benefits of having access to quality blue environments like wetlands? (Para. C)

4. What challenges are still critical despite the fact that nature-related solutions concerning wetland have risen lately? (Para. E)

5. What is of great importance to enable the internationally established mechanisms and instruments to protect ecosystems effectively? (Para. F)

Exercise 3

Directions: Read the passage again and write a summary of **Passage 2** with no more than 150 words.

Extensive Reading and Writing

Directions: For this part, you are required to read the following passage and then write an essay on **The Importance of Wetlands to Mankind**. You should write at least 120 words but no more than 180 words.

Wetland Status, Trends and Response

Royal C., Gardner, C. Max Finlayson

Conservation and wise use of wetlands are vital for human livelihoods. The wide range of ecosystem services wetlands provide means that they lie at the heart of sustainable development. Yet policy and decision-makers often underestimate the value of their benefits to nature and humankind. Understanding these values and what is happening to wetlands is critical to ensuring their conservation and wise use. The Global Wetland Outlook summarizes wetland extent, trends, drivers of change and the steps needed to maintain or restore their ecological character.

Accuracy of global wetland area data is increasing. Global inland and coastal wetlands cover over 12.1 million square kilometers, an area almost as large as Greenland, with 54% permanently inundated and 46% seasonally inundated. However, natural wetlands are in long-term decline around the world; between 1970 and 2015, inland and marine/coastal wetlands both declined by approximately 35%, where data are available, three times the rate of forest loss. In contrast, human-made wetlands, largely rice paddy and reservoirs, almost doubled over this period, now forming 12% of wetlands. These increases have not compensated for natural wetland loss.

Biodiversity

Overall available data suggest that wetland dependent species such as fish, waterbirds and turtles are in serious decline, with one-quarter threatened with extinction particularly in the tropics. Since 1970, 81% of inland wetland species populations and 36% of coastal and marine species have declined. Global threat levels are high (over 10% of species globally threatened) for almost all inland and coastal wetland-dependent taxa assessed. Highest levels of extinction threat (over 30% of species globally threatened) are for marine turtles, wetland-dependent megafauna, freshwater reptiles, amphibians, non-marine molluscs, corals, crabs and crayfish. Extinction risk appears to be increasing. Although waterbird species have a relatively low global threat level, most populations are in long-term decline. Only coral reef-dependent parrotfish and surgeonfish, and dragonflies have a low threat status.

Water quality

Water quality trends are mostly negative. Since the 1990s, water pollution has worsened in almost all rivers in Latin America, Africa and Asia. Deterioration is projected to escalate. Major threats include untreated wastewater, industrial waste, agricultural runoff, erosion and changes in sediment. By 2050, one third of the global population will likely be exposed to water with excessive nitrogen and phosphorous, leading to rapid algal growth and decay that can kill fish and other species. Severe pathogen pollution affects one-third of rivers in Latin America, Africa and Asia, with faecal coliform bacteria increasing over the last two decades. Salinity has built up in many wetlands, including in groundwater, damaging agriculture. Nitrogen oxides from fossil fuels and ammonia from agriculture cause acid deposition. Acid mine drainage is a major pollutant. Thermal pollution from power plants and industry decreases oxygen, alters food chains and reduces biodiversity. At least 5.25 trillion persistent plastic particles are afloat in the world's oceans and have huge impacts in coastal waters. In nearly half OECD countries, water in agricultural areas contains pesticides above national recommended limits. These impacts harm our health, undermine ecosystem services and further damage biodiversity.

Ecosystem processes

Wetlands are one of the most biologically productive ecosystems. They play a major role in the water cycle by receiving, storing and releasing water, regulating flows and supporting life. River channels, floodplains and connected wetlands play significant roles in hydrology, but many "geographically isolated" wetlands are also important. However, land use change and water regulation infrastructure have reduced connectivity in many river systems and with floodplain wetlands. Wetlands regulate nutrient and trace metal cycles and can filter these and other pollutants. They store the majority of global soil carbon, but in the future climate change may cause them to become carbon sources, particularly in permafrost regions.

Ecosystem services

Wetland ecosystem services far exceed those of terrestrial ecosystems. They provide critical food supplies including rice and freshwater and coastal fish, and fresh water, fibre and fuel. Regulating services influence climate and hydrological regimes, and reduce both pollution and disaster risk. Natural features of wetlands often have cultural and spiritual importance.

Urgent action is needed at the international and national level to raise awareness of the benefits of wetlands, put in place greater safeguards for their survival and ensure their inclusion in national development plans. In particular:

- Enhance the network of Ramsar Sites and other wetland protected areas: designation of over 2,300 internationally important wetlands as Ramsar Sites is encouraging. However, designation is not enough. Management plans must be developed and implemented to ensure their effectiveness. Less than half Ramsar Sites have done this as yet.
- Integrate wetlands into planning and the implementation of the post-2015 development agenda: include wetlands in wider scale development planning and action including the Sustainable Development Goals, the Paris Agreement on Climate Change and the Sendai Framework on Disaster Risk Reduction.
- Strengthen legal and policy arrangements to protect all wetlands: wetland laws and policies should apply cross-sectorally at every level. National Wetland Policies are needed by all countries. An important tool here is the avoid–mitigate–compensate sequence recommended by Ramsar and reflected in many national laws. It is easier to avoid wetland impacts than to restore wetlands.
- Implement Ramsar guidance to achieve wise use: Ramsar has a wide range of relevant guidance. Ramsar mechanisms—such as reports on changes in ecological character, the Montreux Record of Ramsar Sites at risk and Ramsar Advisory Missions—help to identify and address challenges to the conservation and management of Ramsar Sites.
- Apply economic and financial incentives for communities and businesses: funding for wetland conservation is available through multiple mechanisms, including climate change response strategies and payment for ecosystem services schemes. Eliminating perverse incentives has positive benefits. Businesses can be helped to conserve wetlands through tax, certification and corporate social responsibility programmes. Government investment is also critically important.
- Integrate diverse perspectives into wetland management: multiple wetland values must be taken into account. To ensure sound decision-making, stakeholders need an understanding of wetland ecosystem services and their importance for livelihoods and human well-being.
- Improve national wetland inventories and track wetland extent: knowledge supports innovative approaches to wetland conservation and wise use. Examples include remote sensing and field assessments, citizen science and incorporating indigenous and local knowledge. Identification and measurement of indicators of wetland benefits and drivers of change are key to supporting wise use policy and adaptive management.

A broad range of effective wetland conservation options is available at the international, national, catchment and site level. Good governance and public participation are critical throughout, management is required, investment essential and knowledge critical.

Self-assessment

Read the self-assessment guidelines for *Self-assessment scale for reading comprehension* and *Self-assessment scale for written expression* (based on the *China's Standards of English Language Ability*) outlined in the appendix of this textbook. Please conduct a self-assessment to evaluate your own skills respectively.

Unit 5 Grassland Ecosystem

Learning Objectives

After learning this unit, you will be able to:

- Understand the mechanisms that affect grassland biodiversity and enhance perceptions of grassland degradation.
- Learn how to read tables and figures in academic context.
- Select proper visuals for data display and describe visuals in academic essays.

Part I Reading for Academic Purposes

Reading Tables and Figures

Visual elements like tables and figures are frequently found to display quantitative information as evidence for the argument in academic texts. They are regarded as the quickest way to help readers understand large quantities of data that would be complicated to explain in text.

Tables are defined by rows and columns containing text or numerical data. Figures are defined as any visual element that is not a table. Figures can take many forms, such as graphs, charts, diagrams, photographs, graphics, maps, etc.

Each table or figure needs a clear and informative legend or caption, consisting of a descriptive title, functioning as the "topic sentence" of the visual, and the explanatory text accompanying the table or figure. Legends can be lengthy or short, depending on the discipline. Legends for tables are usually placed above the table, and legends for figures are placed below the figure.

Tables and figures are typically numbered and presented sequentially and referred to in that order in the text by "Table 1", "Figure 1 (Fig.1)", etc. If the visuals have been copied or adapted from sources of information, they need to be correctly labelled and referenced.

Although tables and figures are mostly aimed at complementing information in the text, they must make sense alone. This means the reader does not need to rely on the text of the article to understand the data presented in the table or figure, and all necessary information are expected to be included in the table/figure, in the legend/caption, or in keys or footnotes.

For example, if readers are interested in the experiment reported in the following **Passage 1**, they can just jump to Figure 1 to learn about the experiment treatment and results, which are illustrated distinctly and vividly by means of pictures and specified legend. For the relevant text referring to the figure in the passage, the author not only highlights the most important findings, but provides relevant analysis and interpretation of those findings.

Reading Passage 1

Critical Thinking Questions

Directions: Read the following passage and answer the questions.

1. What was the purpose of Eskelinen and colleagues' field experiment?
2. Which three factors are highlighted to explain diversity declines in grassland?
3. What are the implications of the experiment?

Shedding Light on Declines in Grassland Biodiversity

Eric Allan

A. How different plant species coexist is a central question in ecology. In an era of global change and rapid species loss, understanding the **mechanisms** that affect biodiversity is a matter of urgency. Research shows that adding nutrients to the soil through the use of fertilizer, or removing large **grazing** animals, such as cattle and sheep, reduces the diversity of plant species in grasslands. Plants generally grow taller when nutrients are added or when grazers are removed, and these tall, fast-growing plants are thought to reduce diversity because they can **monopolize** access to light and **shade** out smaller plants.

B. However, until now, only indirect evidence was available to support the **hypothesis** that a rise in plant competition for light was responsible for declines in diversity. Supporting data include, for example, studies relating changes in light levels to diversity changes after removal of grazers, or analyses that use increases in plant biomass as a way to estimate light changes. Eskelinen et al. conducted an experiment to gather direct evidence by testing the effect of plant competition for light at a German grassland site.

C. Eskelinen and colleagues placed lamps in the vegetation to increase the light levels for short species (Figure 1). The authors report that this light addition **reversed** most of the diversity decline caused by the removal of grazers and, in the short term, reduced the diversity loss caused by adding fertilizer. This result supports the hypothesis that grazing animals increase the amount of light available to plants by keeping vegetation low and thereby prevent tall plant species from capturing all the light. On the **plots** without sheep, diversity was lost relative to plots with grazers, and plant cover and **decaying** plant-litter material increased, as expected. Such litter might **suppress** seedling growth and prevent less-**robust** species from **colonizing** such areas. But when lamps were shone on these plots, the decline in species diversity was **substantially** reduced, compared with the plots without grazers that did not receive a light **boost**. It therefore seems probable that grazers promote plant **coexistence**, principally by **equalizing** the ability of species to compete for light.

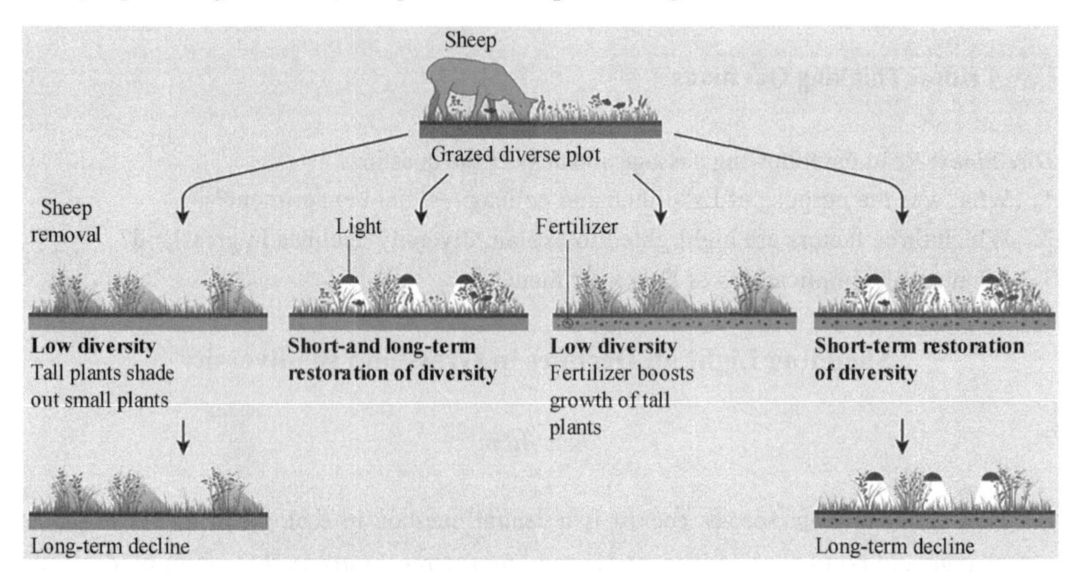

Figure 1 How light, grazing and fertilizer use affect grassland diversity. Eskelinen et al. carried out a field experiment to investigate biodiversity changes driven by plant competition. Grazing sheep boost diversity by preventing tall plants from becoming too high and thus shading out small plants (in the understorey). The authors provide direct evidence that this is a light-dependent mechanism, because when the grazers were removed, diversity was restored in the short and long term by providing light in the understorey. The addition of fertilizer lowers diversity by boosting the growth of tall plants, and such a fertilizer-driven decline in diversity was prevented by light supplementation in the short term, but not in the long term. The latter result indicates that unknown, light-independent factors contribute to diversity declines associated with long-term fertilizer use.

D. The major plants that benefited from higher light levels were those with a slow and conservative strategy for resource use. These slow-growing plants would otherwise **lose out to** faster-growing, "greedy" species that efficiently capture light and, by comparison, do not invest much in other aspects of plant development, such as defence. This work is the first direct experimental demonstration that plant competition for light explains the diversity declines observed when grazers are removed.

E. The experiment consisted initially of only a "cross" of fertilizer and light addition on grassland plots where sheep had just been removed—the plots contained all four possible combinations of addition or lack of **supplementation** of light and fertilizer. After a few months had **elapsed**, Eskelinen and colleagues could already observe a decline in plant diversity where they added fertilizer, indicating that a decrease in light levels for low-growing plants, as a consequence of fertilizer boosting the tall plants, was driving the diversity decline. However, this decline was **halted** when light was shone on the plants growing close to the ground in the **understorey**, consistent with findings from a greenhouse experiment carried out under more-controlled conditions.

F. In the second year of the study, Eskelinen and colleagues let sheep graze again on an extra set of plots, to compare what happened when fertilizer and light were introduced to grazed or ungrazed plots. **Intriguingly**, by the third year of the study, light addition no longer prevented diversity declines when fertilizer was added. Grazing had such a dominant effect on the available light levels for low-growing plants that it **overwhelmed** the effect of fertilizers—light levels were always high on grazed plots regardless of **fertilization**, and on ungrazed plots, light levels in the understorey were so low that adding fertilizer did not decrease understorey light further.

G. However, this is not the whole story: adding fertilizer still caused a decline in diversity in both grazed and ungrazed plots. This diversity decline must have been driven by other light-independent mechanisms, such as an increase in litter that reduced seedling success, or because high nutrient levels drove the loss of nutrient **niches** that would have favoured the coexistence of plants with distinct abilities to capture particular limiting nutrients. So, although changes in plant competition for light can explain some of the effects of fertilization on diversity, other mechanisms must have a role, too. The experiment raises the interesting possibility that the relative contribution of light limitation on biodiversity varies over time, or changes as communities reassemble.

H. Eskelinen and colleagues' study is extremely valuable because it experimentally **manipulates** light in the field and clearly shows that changes in light levels explain the effect that **herbivore** grazing has on diversity. The work also raises many questions. It is unclear why light addition was unable to boost diversity on fertilized plots in the later years of the study. Is it the case that climate-induced variation in plant biomass production shifts the relative strength of competition for light from year to year? However, without multi-year data sets and information on plant productivity, it is not possible to test this idea. **Alternatively**, does fertilization result in reduced light levels at ground level only immediately after herbivores are removed, after which light limitation becomes less central for explaining the effects of fertilizer addition? At some point after the first year, the effects of sheep grazing on light availability became so strong that adding fertilizer did not alter light levels for low-growing plants. Unfortunately, data were not collected during the second year of the experiment, so the authors could not examine at what point grazer removal overwhelms the effect of fertilizer on light in the understorey.

I. Questions aside, the study's results highlight the value and power of this type of field experiment, in which several treatments are crossed with each other. The dominant effect of the grazers on diversity and on the availability of light underlines the crucial role of herbivores in structuring plant communities and maintaining diversity,

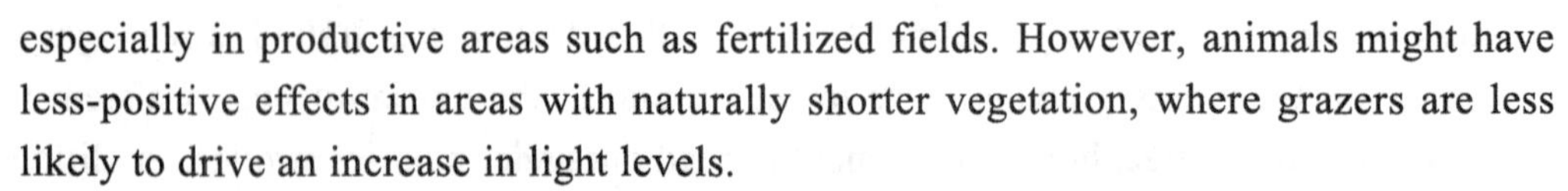
especially in productive areas such as fertilized fields. However, animals might have less-positive effects in areas with naturally shorter vegetation, where grazers are less likely to drive an increase in light levels.

J. It would be interesting to extend the ideas explored here, by testing whether **invertebrate** herbivores or disease-causing fungi might also promote diversity by causing a rise in light levels. Eskelinen and colleagues' study demonstrates how a simple experimental set-up can be used to **pinpoint** mechanisms underlying declines in diversity. More such investigations are needed to reveal fundamental mechanisms underlying plant coexistence and to improve our ability to predict how global change will alter biodiversity.

Key Words and Phrases

1. shed light on		to provide new information that makes a difficult subject or problem easier to understand 阐明，使更易于理解
2. mechanism /'mekənɪzəm/	*n.*	the way that sth. works 机制；原理
3. graze /greɪz/	*v.*	(of cows, sheep, etc.) to eat grass that is growing in a field 放牧；吃草
4. monopolize /mə'nɒpəlaɪz/	*v.*	to have complete control over something so that other people cannot share it or take part in it 独占；垄断
5. shade /ʃeɪd/	*v.*	to prevent direct light from shining on something 遮（光）
6. hypothesis /haɪ'pɒθəsɪs/	*n.*	an idea or explanation for something that is based on known facts but has not yet been proved 假设，假说
7. reverse /rɪ'vɜːs/	*v.*	to change something completely so that it is the opposite of what it was before（使）反向；（使）倒转
8. plot /plɒt/	*n.*	a small piece of land for growing things on 小块土地
9. decay /dɪ'keɪ/	*v.*	to be destroyed gradually by natural processes（使）腐烂
10. suppress /sə'pres/	*v.*	to prevent sth. from growing, developing or continuing 抑制；阻止……生长
11. robust /rəʊ'bʌst/	*adj.*	strong and healthy 强壮的，茁壮的
12. colonize /'kɒlənaɪz/	*v.*	(of animals or plants) to live or grow in large numbers in a particular area（动植物）在（某一地区）大批生长
13. substantially /səb'stænʃəli/	*adv.*	very much; a lot 非常；大大地

14.	boost /buːst/	*n.*	an increase in something 增加，增长
15.	coexistence /ˌkəʊɪg'zɪstəns/	*n.*	the state of being together in the same place at the same time 共存，共处
16.	equalize /'iːkwəlaɪz/	*v.*	to make two or more things the same in size, value, amount etc.使均等；使平等
17.	lose out to someone		to lose a competition to someone 输给某人
18.	supplementation /ˌsʌplɪmen'teɪʃn/	*n.*	the act of adding something to something else in order to improve or complete it 增补；补充；追加
19.	elapse /i'læps/	*v.*	(time) pass, go by（时间）流逝，过去
20.	halt /hɔːlt/	*v.*	to (cause to) stop moving or doing something or happening（使）停止，停下
21.	understorey /ʌndərs'tɔːrɪ/	*n.*	an underlying layer of vegetation 下层植被
22.	intriguingly /ɪn'triːgɪŋli/	*adv.*	in a way that is very interesting because of being unusual or mysterious 非常有趣地
23.	overwhelm /ˌəʊvə'welm/	*v.*	to overcome by superior force or numbers 压倒，战胜
24.	fertilization /ˌfɜːtəlaɪ'zeɪʃ(ə)n/	*n.*	the act of adding a substance to soil to make plants grow more successfully 施肥
25.	niche /niːʃ/or/nɪtʃ/	*n.*	(ecology) the status of an organism within its environment and community (affecting its survival as a species) 生态位，小生境
26.	manipulate /mə'nɪpjuleɪt/	*v.*	to handle, manage, or use, especially with skill, in some process of treatment or performance 操作；控制
27.	herbivore /'hɜːbɪvɔː(r)/	*n.*	an animal that eats only plants 草食动物
28.	alternatively /ɔːl'tɜːnətɪvli/	*adv.*	used to suggest another possibility 要不，或者
29.	invertebrate /ɪn'vɜːtɪbrət/	*adj.*	(of an animal) without a backbone 无脊椎的
30.	pinpoint /'pɪnpɔɪnt/	*v.*	to discover or describe the exact facts about sth. 查明；准确描述

Reading Comprehension

Directions: Read **Passage 1** again and answer the following questions.

1. In Paras. A—C, which is NOT true about the role of grazing in affecting grassland diversity?

A. Removing grazers allows tall plants to monopolize access to light.

B. Grazing keeps vegetation low, increasing the light levels for short species.

C. Grazing animals may suppress seedling growth, reducing diversity.

D. Grazers promote plant coexistence, equalizing light availability of species.

2. According to Para. F, by the third year of the study, which measure seemed to be the most effective in preventing diversity declines?

A. Grazing. B. Fertilization.

C. Light addition. D. Non-grazing.

3. According to Para. H, what can you learn from Eskelinen and colleagues' study?

A. Evidence for the short-term restoration of biodiversity with light addition is available.

B. Variations in light levels explain the effect of grazing on diversity.

C. Collecting data during the second year of the experiment met great difficulty.

D. Only light limitation can explain the effects of fertilizer addition.

4. Which aspect is NOT proposed concerning future research on biodiversity declines?

A. Effects of invertebrate herbivores.

B. Effects of disease-causing fungi.

C. Mechanisms underlying plant coexistence.

D. Process of desertification.

Vocabulary Exercises

Exercise 1

Directions: In this section, there are ten sentences, each one with one word missing. You are required to complete these sentences with the proper form of the words given in the brackets.

1. That fact raises some ____________ possibilities about what we might discover beyond the bounds of Earth. (intrigue)

2. As well, qualitative research is frequently used in psychology as a way of generating informed ____________ for a given topic. (hypothesize)

3. To upgrade common service in ethnic area and to ____________ it are necessities to consolidate and develop the socialism national relationships. (equal)

4. The quickest way is to take the motorway. ____________, there is the coastal route, which is much prettier. (alternative)

5. The United States, and really, the entire world, has squandered much of the time that has ____________ since climate change first became a concern more than forty years ago. (elapse)

6. It is unclear why light addition was unable to boost diversity on ____________ plots in the later years of the study. (fertile)

7. Yet no one has ever been able to ____________ any of the innovations that have periodically roiled the game. (monopoly)

8. There were two considerations that made the ____________ of the sale of beer difficult for the colonial authorities. (suppress)

9. With vast deserts stretching north and south of the equator, Kent says, there would have been few plants available for the ____________ to survive the journey north for much of that time period. (herbivore)

10. The authors report that this light addition ____________ most of the diversity decline caused by the removal of grazers. (reverse)

Exercise 2

Directions: In this section, there are ten sentences with ten blanks. You are required to select one word for each blank from a list of choices given in a word bank. Each choice in the bank is identified by a letter. You may not use any of the words in the bank more than once.

A. boost	B. substantially	C. shade	D. reassembled
E. understorey	F. niche	G. overwhelmed	H. principally
I. mechanism	J. supplementation	K. efficiently	L. conservative
M. initial	N. decay	O. robust	

1. Without relative economic strength, the sphere's political and military strength will ____________ over time.

2. Trees provide homes for animals and ____________ for people on hot days.

3. Given how fast and widely the virus is spreading, those risks are going to be ____________ higher by Thanksgiving.

4. El Nino is known to ____________ the planet's temperature, as warmth stored within the Pacific Ocean pools at its surface along the equator, driving changes in weather patterns around the world and releasing more heat and humidity into the atmosphere.

5. We recorded a total of 88 ____________ species, 62 of which were non-woody species and 26 were woody species.

6. Will tropical forests also be ____________ by exotic species, or do tropical forests possess some resistance against invasive species?

7. The bricks were numbered so that it could be easily and correctly ____________ on the preservation site.

8. According to a paper released in March, the hippos might actually be filling an ecological ____________ that's been empty for tens of thousands of years.

9. Worthwhile results obtained from this framework must be ___________ enough to be independent of small numerical differences, which may be without real significance.

10. There is also no solid evidence that Vitamin D ___________ will lower the risk of becoming infected or developing a serious case of COVID-19.

Exercise 3

Directions: Read the passage again and translate the following sentences from **Passage 1** (underlined in the passage) into Chinese.

1. The experiment consisted initially of only a "cross" of fertilizer and light addition on grassland plots where sheep had just been removed—the plots contained all four possible combinations of addition or lack of supplementation of light and fertilizer.

2. Grazing had such a dominant effect on the available light levels for low-growing plants that it overwhelmed the effect of fertilizers—light levels were always high on grazed plots regardless of fertilization, and on ungrazed plots, light levels in the understorey were so low that adding fertilizer did not decrease understorey light further.

3. This diversity decline must have been driven by other light-independent mechanisms, such as an increase in litter that reduced seedling success, or because high nutrient levels drove the loss of nutrient niches that would have favoured the coexistence of plants with distinct abilities to capture particular limiting nutrients.

4. Alternatively, does fertilization result in reduced light levels at ground level only immediately after herbivores are removed, after which light limitation becomes less central for explaining the effects of fertilizer addition?

Part II Academic Writing Strategies

Using Visuals in Academic Writing

Since visuals contain many types, selecting the right kind of visual to convey the research story is crucial. Table 6 summarizes some common types of visuals and their uses.

Table 6 Common Types of Visuals in Academic Writing

<table>
<tr><th>Type of Visuals</th><th>Example</th><th>Uses</th></tr>
<tr><td>Table</td><td><table>
<tr><th></th><th>Adidas</th><th>Nike</th><th>Anta</th><th>Umbro</th><th>Reebok</th></tr>
<tr><td>Jan</td><td>555</td><td>278</td><td>182</td><td>408</td><td>584</td></tr>
<tr><td>Feb</td><td>303</td><td>402</td><td>522</td><td>245</td><td>358</td></tr>
<tr><td>Mar</td><td>424</td><td>163</td><td>364</td><td>158</td><td>852</td></tr>
<tr><td>Apr</td><td>420</td><td>357</td><td>170</td><td>952</td><td>752</td></tr>
</table></td><td>Place detailed data/information in categories formatted into rows and columns for comparison; show the actual data values and their precision</td></tr>
</table>

continued table

Type of Visuals	Example	Uses
Column and bar graph/chart	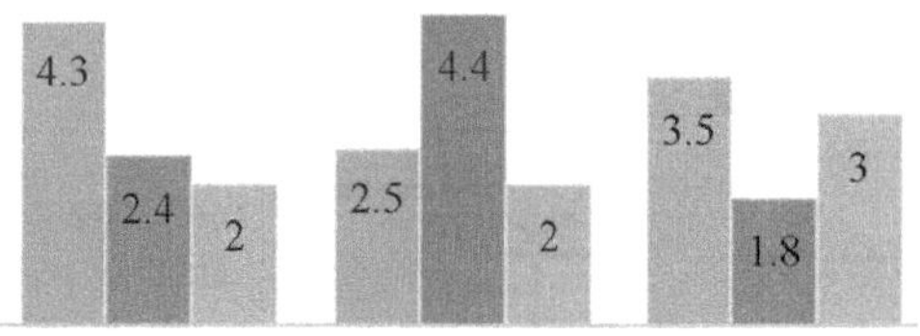	Compare and contrast two or more subjects at the same point in time; compare change over time
Line graph/chart	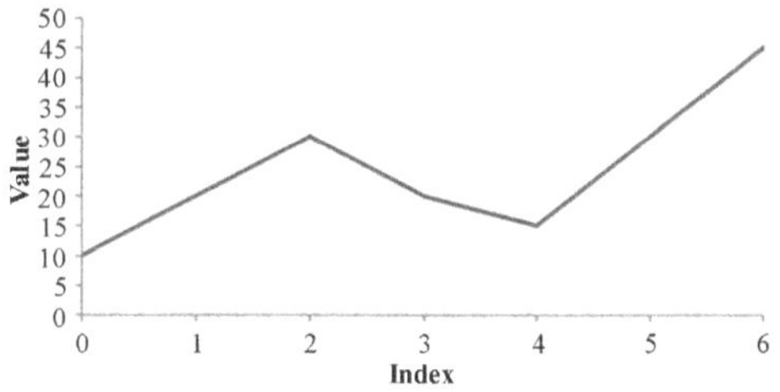	Show a sequence of variables in time or space; display other dependent relationships (e.g. change over time)
Pie chart	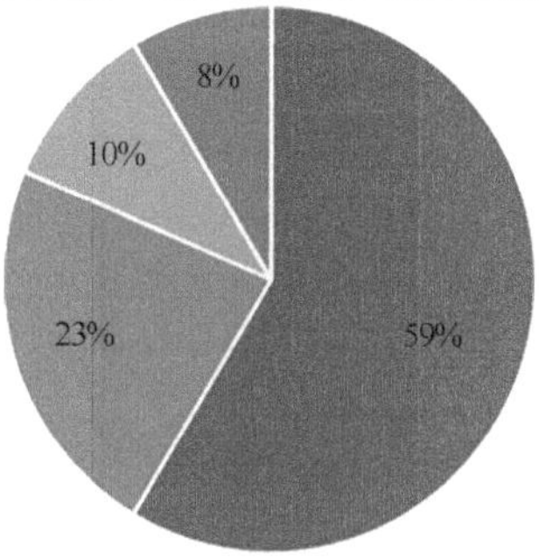	Display the number and relative size of the divisions of a subject; show relation of parts to a whole (parts must sum to 100% to make sense)
Flow chart	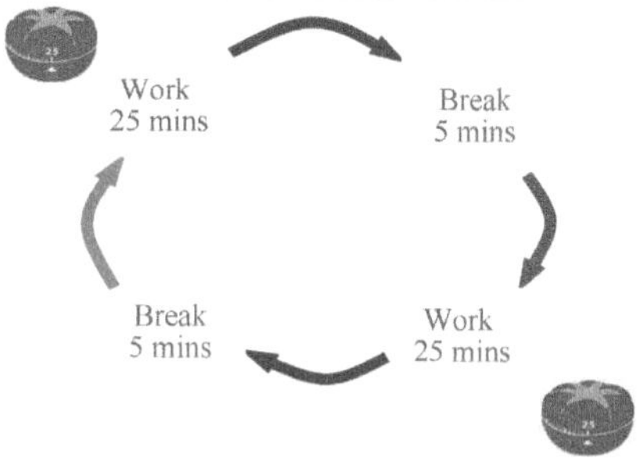	Show the sequence of steps in a process or procedure
Gantt chart	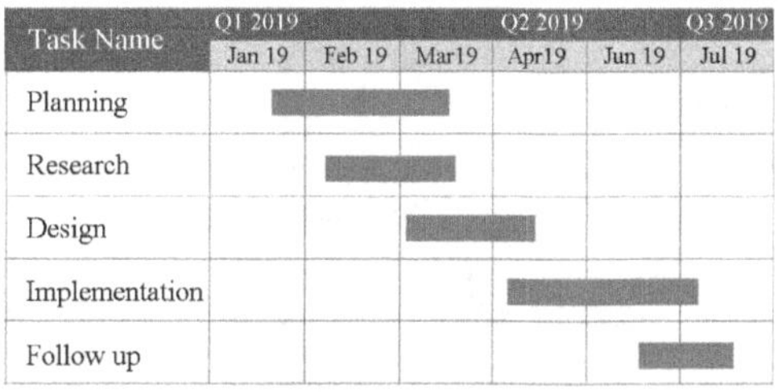	Indicate timelines for multi-stepped projects, especially used in proposals and progress reports

continued table

Type of Visuals	Example	Uses
Diagram	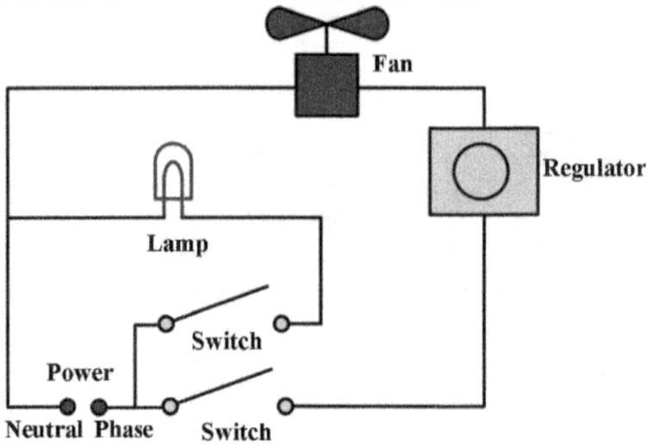Fan Regulator Lamp Switch Power Neutral Phase Switch	Display the parts of a subject and their spatial or functional relationship; emphasize details; show dimensions
Scatterplot (X-Y Graph)	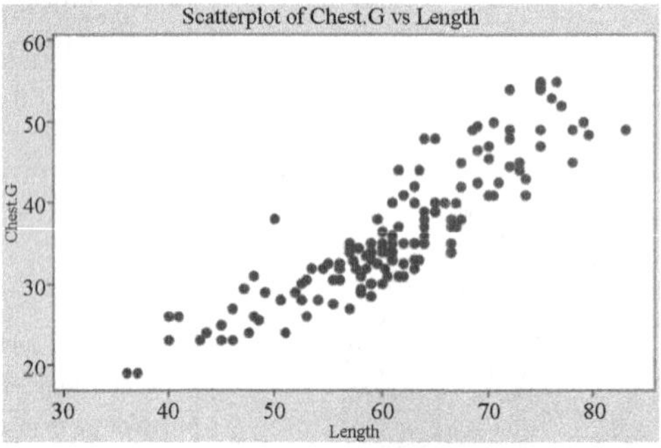Scatterplot of Chest.G vs Length 60 50 40 30 20 Chest.G 30 40 50 60 70 80 Length	Show the relationship between two or more quantitative variables; show trends or relationships in the data over time; plot the independent variable on the X axis (the horizontal axis) and the dependent variable on the Y axis (the vertical axis)
Radar chart	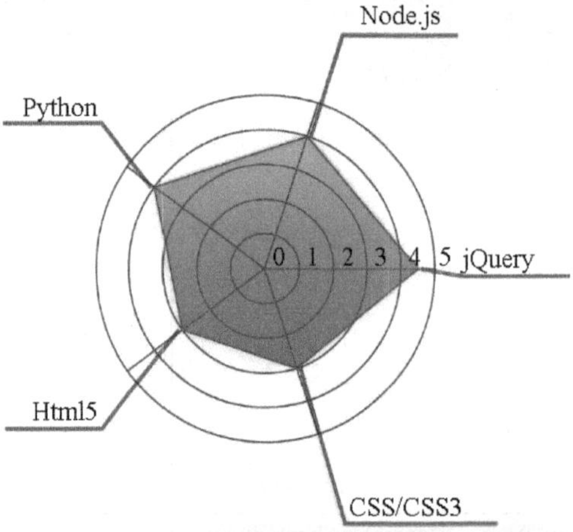Node.js Python 0 1 2 3 4 5 jQuery Html5 CSS/CSS3	Compare three or more than three variables when categories are not directly comparable
Photo	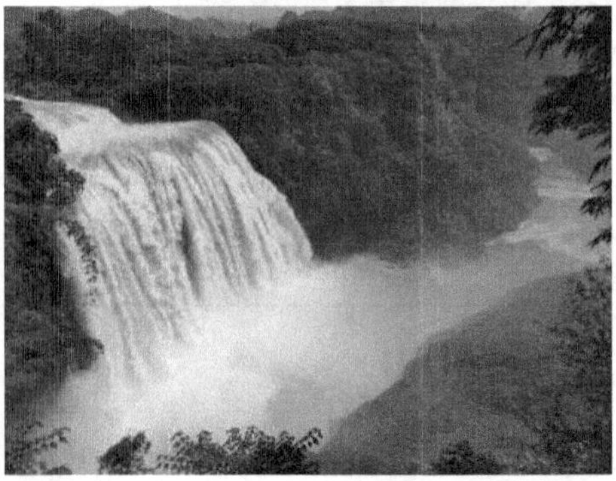	Show what a subject looks like in realistic detail

continued table

Type of Visuals	Example	Uses
Map		Show the location of the study site

Before inserting visuals in an academic essay, the author should clarify the relations between the visuals and the relevant text. As is stated in the reading part, effective visuals simplify complicated textual descriptions and help the reader understand a complicated process or visualize trends in the data. All visuals have to be based on and supplement the written text rather than duplicate it. Likewise, when describing visuals in text, it is unnecessary to repeat all data from the visuals. Focus of text should be: highlighting the main or important findings, showing relationships between data, and summarizing information or trend in the table or figure, etc.

Every visual used must be referred to in the text. The reference to the figure or table should be informational and draw the reader's attention to the relationship or trend being highlighted.

Here are some ways of referring to data in a table or figure.

(1) Refer to the table or figure in parenthesis in the middle or at the end of the sentence.

Although light addition reduced the loss of species richness (Figure 2), the interaction between light addition and fertilization was weaker on species richness than on Shannon diversity.

(2) Use "as", either at the beginning or end of the sentence.

As shown in Table 3, herbivore exclusion decreased species richness by 12.5% and Shannon diversity by 11.7%.

Herbivore exclusion decreased species richness by 12.5% and Shannon diversity by 11.7%, as shown in Table 3.

(3) Start with table/figure number and make a statement with it.

Figure 1 indicates that below-ground carbon inputs tend to be more often incorporated into SOC than above-ground inputs.

(4) Use passive voice.

Consistent reduction of soil carbon stocks by moderate and heavy (continuous) grazing is shown in Table 2.

Reading Passage 2

Critical Thinking Questions

Directions: Read the following passage and answer the questions.

1. According to the authors, how is grassland degradation defined?
2. What drivers are mentioned resulting in grassland degradation?
3. What role do woody plants play in grassland degradation?

Extent and Drivers of Grassland Degradation

Richard D. Bardgett, James M. Bullock, Sandra Lavorel, et al.

A. Globally, estimates suggest that as much as 49% of the total grassland and about half of the natural grassland area has been degraded to some extent. However, specific estimates of the extent of grassland degradation vary greatly. For instance, about 90% of the UK's semi-natural species-rich grasslands have been degraded as result of intensive agriculture and land conversion since the 1940s. Estimates suggest that up to 90% of the vast grasslands of the Qinghai–Tibetan Plateau have been degraded to some extent due to human activities and climate change. Over 60% of the former grassland area of southern Brazil has been lost to unsuitable management and land use change.

B. A key issue when considering grassland degradation is clarifying what it **constitutes**, especially since definitions of both grassland and degradation vary. Degradation is broadly defined as a decline in land condition, or, more specifically, the human-caused processes that drive a persistent decline or loss in biodiversity, ecosystem functions or ecosystem services. Such a broad definition of grassland degradation can help to inform policy and practice about the level of grassland degradation in a certain region. However, pinpointing what degradation constitutes in specific settings can be less clear because it has both ecological and socio-economic dimensions, and different stakeholder groups prioritize different combinations of ecosystem services. As such, the de- finition of grassland degradation can depend upon the stakeholder and needs to be **tailored** accordingly.

C. Grassland degradation is defined here from a socio-ecological perspective, where grassland is considered degraded if the supply of multiple ecosystem services **falls short of** that demanded by grassland stakeholders. In some cases, grassland degradation is apparent to all grassland stakeholders, for instance, when overgrazing leads to loss of vegetation, declines in soil organic matter and consequent soil erosion, or when natural or semi-natural grassland is converted to another land use. However, in many cases, **perceptions** of degradation could differ: the increase in plant production but accompanying loss of plant species diversity resulting from fertilizer use might be considered an improvement by **pastoralists** (due to increased forage production) but degradation by conservationists concerned with biodiversity protection. Hence, defining grassland degradation from a socio-ecological perspective recognizes that it alters the supply of multiple ecosystem services from grassland relative to their human demand by different stakeholders. Moreover, a socio-ecological approach provides a framework for guiding the restoration of degraded grasslands, as it considers the need to enhance the co-supply of multiple ecosystem services in efforts to meet the needs of all stakeholders.

D. Although threats to natural and semi-natural grasslands are present globally (Figure 2), the tropics face particularly **acute** threats, while degradation of European semi-natural grassland has largely slowed. Human activities are the principal drivers of grassland degradation. For example, increased disturbance from overgrazing or a **heightened** fire frequency has reduced vegetation cover, increasing **susceptibility** to soil erosion and desertification. Conversely, the **cessation** of livestock grazing associated with extensive land abandonment of semi-natural grasslands in Europe during the twentieth century has led to grassland degradation due to the expansion of **scrub**. Nutrient enrichment of natural and semi-natural grasslands from fertilizer use and/or atmospheric nitrogen deposition has led to widespread declines in biodiversity and other ecosystem services. These impacts are especially notable when combined with the sowing of productive **cultivars** to support heavy grazing and/or **silage** production in intensively managed grasslands. The conversion of natural and semi-natural grasslands to other land uses, such as arable farming, built infrastructure and forestry, is also a major driver of grassland degradation worldwide. One immediate land use threat is the planting of trees on natural grasslands, **ostensibly** to meet afforestation targets for climate change mitigation.

E. Ongoing climate change also poses a threat to all grasslands, as it causes grassland degradation. **Projected** future climate change will likely combine with human activities to cause increased woody plant **encroachment** in some areas and desertification in others. For example, natural grasslands in the Americas, Australia and Africa are being degraded due to woody plant encroachment, with the major causes thought to be a combination of higher atmospheric CO_2 concentrations, warmer and wetter conditions, and changes in grazing intensity and timing relative to fire, which is key to **episodic** tree recruitment. Conversely, in many parts of the world, such as Central Asia, overgrazing combined with more intense and frequent droughts is **exacerbating** grassland desertification and degradation.

①

United States
Reseeding and fertilization of tallgrass prairie has transformed species-rich prairie grasslands with a mixture of native C3 and C4 grasses, sedges and forbs to species-poor grass- land dominated by Eurasian C3 grasses.

②

Brazil
Cultivation and abandonment of Cerrado grasslands poses a major threat to these ancient and highly diverse ecosystems.

③

United Kingdom
Ploughing, reseeding and fertilization have transformed species-rich chalk grasslands into "improved" grasslands with higher fertility, lower species richness and lower levels of many cultural ecosystem services.

④

Kenya
Decades of overgrazing by cattle and sheep of lowland grassland in Kenya has caused excessive soil erosion and loss of biodiversity and ecosystem services.

⑤

India
Invasion by exotic woody species, primarily *Acacia mearnsii*, poses a major threat to ancient shola-grassland mosaics in the upper reaches of the Nilgiri Biosphere Reserve, Western Ghats.

⑥

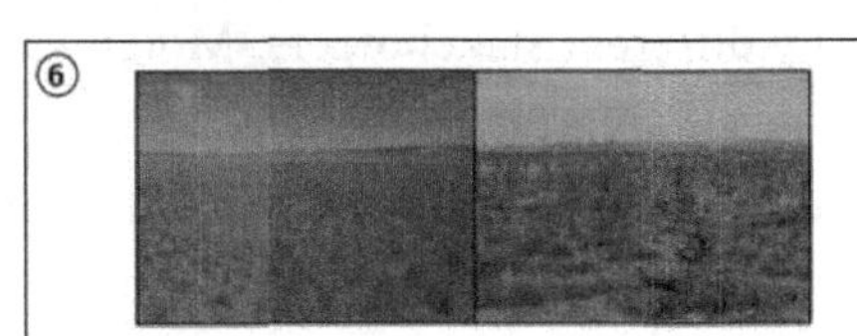

Inner Mongolia, China
Planting of pine trees in semi-arid grasslands can reduce plant diversity, damages soils and increase water shortages.

⑦

Qinghai-Tibetan Plateau, China
Extensive overgrazing and an increase in rodent populations have caused widespread degradation of alpine grasslands, causing soil erosion and loss of biodiversity and ecosystem services.

⑧

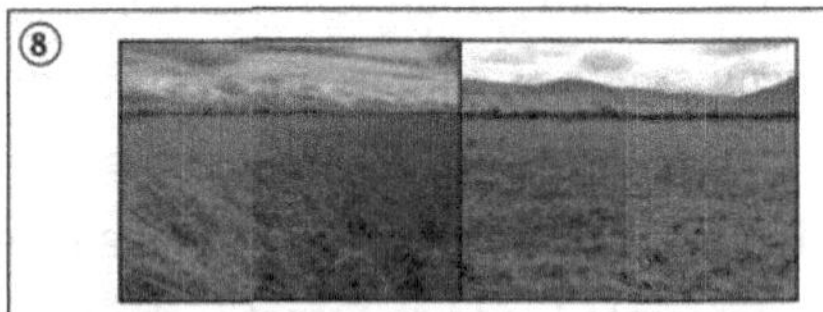

Australia
Livestock grazing has broken down *Themeda* swards and promoted invasion by exotic species.

Figure 2 Degraded grasslands. The extent of degraded grasslands worldwide, with examples of paired non-degraded (left) and degraded (right) grasslands. Grassland classification follows the UN FAO Land Cover Classification System (LCCS) (data downloaded at https://lcviewer.vito.be/2015 with tundra ecosystems excluded). Degradation is measured as greenness changes, as measured by rain use efficiency (RUE) adjusted sum normalized differential vegetation index (NDVI) between 1981 and 2015, with regions showing a reduction in greenness of 0.01 being classed as degraded. Therefore, much degradation involving vegetation change is not shown. Degradation is caused by many factors, including overgrazing, fertilization, tree planting and invasive species. Image of the United States courtesy of L. Brudvig. Image of United Kingdom courtesy of L. Hulmes. Right-hand image of India courtesy of S. K. Chengappa. Image of Australia courtesy of S. Prober.

Key Words and Phrases

1. constitute /'kɒnstɪtjuːt/ *v.* make up, form, compose 组成，构成；制定

2.	tailor /'teɪlə(r)/	*v.*	to make or adapt something for a particular purpose, a particular person, etc. 调整使适应；使合适
3.	fall short of		to fail to reach the standard that you expected or need 没有达到；未能满足；缺乏
4.	perception /pə'sepʃn/	*n.*	a belief or opinion, often held by many people and based on how things seem 认识，观念；感觉；感知
5.	pastoralist /'pɑːstərəlɪst/	*n.*	a farmer who breeds and takes care of animals, especially in Africa and Australia（尤指非洲和澳大利亚的）牧民，牧场主
6.	acute /ə'kjuːt/	*adj.*	very serious or severe 十分严重的
7.	heighten /'haɪtn/	*v.*	to increase the degree or amount of; augment 增加；增强，加强
8.	susceptibility /səˌseptə'bɪləti/	*n.*	the fact that someone or something can easily be influenced, harmed, or infected 敏感性，易受影响（或伤害、感染）
9.	cessation /se'seɪʃn/	*n.*	ending or stopping 结束，停止；中断，中止
10.	scrub /skrʌb/	*n.*	small bushes and trees 矮小的树木，灌木
11.	cultivar /'kʌltɪvɑː(r)/	*n.*	a type of plant that has been deliberately developed to have particular features（植物）栽培品种
12.	silage /'saɪlɪdʒ/	*n.*	grass or other green crops that are stored without being dried and are used to feed farm animals in winter 青贮饲料
13.	ostensibly /ɒ'stensəbli/	*adv.*	according to what seems or is stated to be real or true, when this is perhaps not the case 表面上
14.	projected /prə'dʒektɪd/	*adj.*	estimated for the future or calculated based on information already known 预计的，推断的
15.	encroachment /ɪn'krəʊtʃmənt/	*n.*	the act of slowly covering more and more of an area 侵蚀，侵占
16.	episodic /ˌepɪ'sɒdɪk/	*adj.*	happening only sometimes and not regularly 偶尔发生的；不定期的
17.	exacerbate /ɪg'zæsəbeɪt/	*v.*	to make something that is already bad even worse 使恶化；使加重；使加剧

Writing Tasks

Exercise 1

Directions: Read the following sentences from **Passage 2** and rewrite the following sentences (underlined in the passage), replacing the underlined parts with your own words.

1. Grassland is considered degraded if the supply of multiple ecosystem services falls short of that demanded by grassland stakeholders.

2. Increased disturbance from overgrazing or a heightened fire frequency has reduced vegetation cover, increasing susceptibility to soil erosion and desertification.

3. One immediate land use threat is the planting of trees on natural grasslands, ostensibly to meet afforestation targets for climate change mitigation.

4. Projected future climate change will likely combine with human activities to cause increased woody plant encroachment in some areas and desertification in others.

Exercise 2

Directions: Read **Passage 2** again and answer the questions below. Please give brief answers in about 10 words.

1. Why do authors quote plenty of data at the beginning of the article? (Para. A)
2. What's the primary concern when people discuss grassland degradation? (Para. B)
3. According to the authors, which method is conducive to guiding the restoration of degraded grasslands? (Para. C)
4. What are the dominant drivers of grassland degradation? (Para. D)
5. How do you understand "episodic tree recruitment"? (Para. E)

Exercise 3

Directions: Study Figures 3—5 below and complete the following description with words and expressions in the box.

long-term trend	significant losses	extreme weather events
severely disrupted	an overall increase	estimates
vary widely	global temperatures	mostly due to
nearly triple		

A.

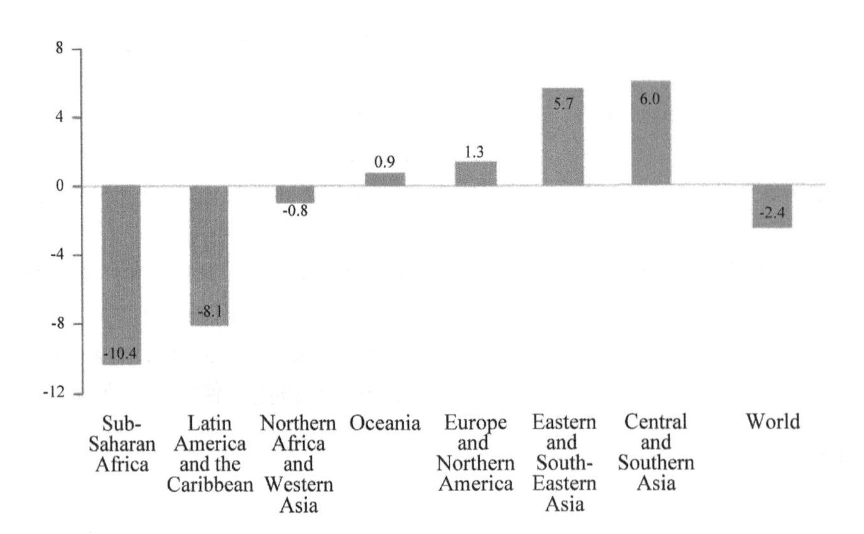

Figure 3 Change of forest area coverage, 2000—2020 (percentage)

Changes in forest area (1) ___________ from region to region. Asia, Europe and Northern America showed (2) ___________ in forest area from 2000 to 2020 due to afforestation, landscape restoration and the natural expansion of forests. In contrast, (3) ___________ were observed in Latin America and sub-Saharan Africa, (4) ___________ the conversion of forests into agricultural land.

B.

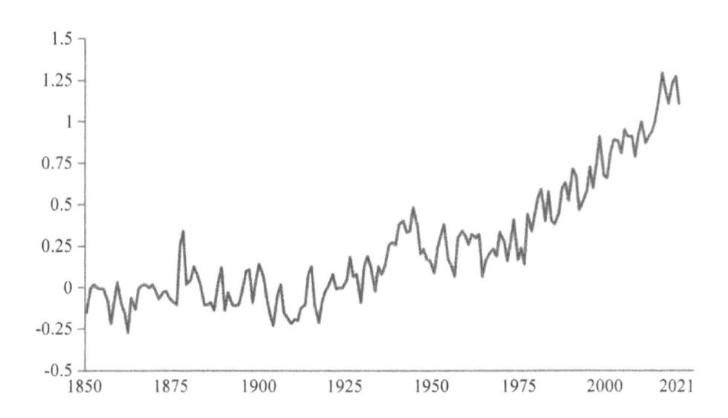

Figure 4 Global annual mean temperature relative to pre-industrial levels (1850—1900 average), 1850—2021 (degrees Celsius)

Source: The figure is drawn from the the World Meteorological Organization's State of the Global Climate 2021 report, which combines six international data sets for temperature: HadCRUT.5.0.1.0 (UK Met Office), NOA-AGlobalTemp v5 (USA), NASA GISTEMP v4 (USA), Berkeley Earth (USA), ERA5 (ECMWF), JRA-55 (Japan).

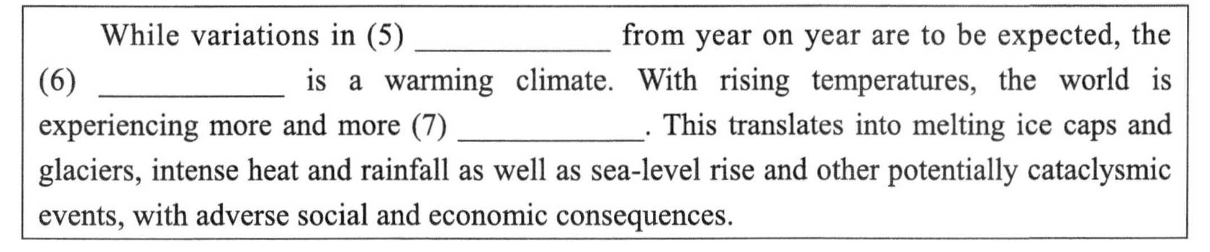

While variations in (5) ___________ from year on year are to be expected, the (6) ___________ is a warming climate. With rising temperatures, the world is experiencing more and more (7) ___________. This translates into melting ice caps and glaciers, intense heat and rainfall as well as sea-level rise and other potentially cataclysmic events, with adverse social and economic consequences.

C.

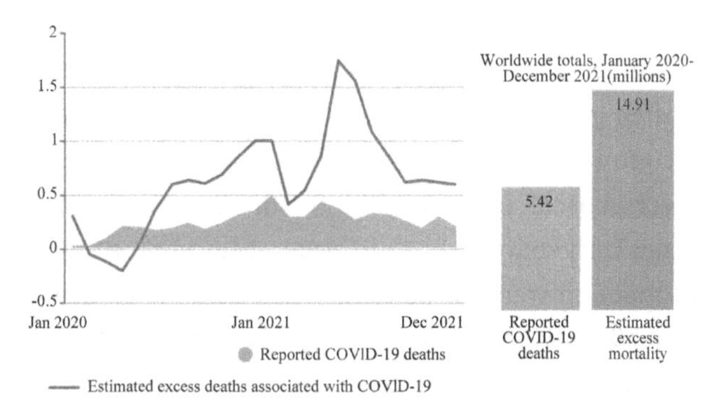

Figure 5 Reported COVID-19 deaths and estimated excess deaths globally, 2020—2021 (millions)

COVID-19 is now a leading cause of death. The latest (8) ____________ suggest that 14.9 million people died as a direct result of COVID-19 or from the pandemic's impact on health systems and society in 2020 and 2021. This estimate is (9) ____________ the 5.4 million officially reported COVID-19 deaths in the same period. The pandemic has (10) ____________ health systems and essential health services.

Exercise 4

Directions: Read the passage again and write a summary of **Passage 2** with no more than 150 words.

Extensive Reading and Writing

Directions: For this part, you are required to read the following passage and then **design a visual (e.g. a mindmap)** to visualize the key information mentioned in this article.

Grasslands: Facts and Information

Loraine Balita-Centeno, Jan Bengtsson, et al.

Grasslands are vast areas covered mostly by grass and grass-like plants. They have many different names. In the US they're called prairies, in Australia, they're called rangelands, while in South America they're called pampas, llanos, and cerrados. These areas don't receive enough rainfall to grow trees or support a forest, but they don't get too little either to become a desert. Grasslands are somewhere in between these two ecosystems. They receive just enough rain to support an abundance of grass and in some areas, a few trees spaced far apart from each other and clusters of shrubs as well. Grasslands support grazing animals or those that mainly feed on the grass to survive like zebra, antelope, and bison.

1. Types of grasslands

There are two main types of grasslands: tropical or savanna and temperate grasslands.

Tropical grasslands. Tropical grasslands are grasslands with scattered trees and some shrubs. Trees often grow away from each other, and in between these widely spaced trees are shrubs that grow in small clumps. These grasslands, also called savannas, are mostly found all over Africa, in Australia, South America, and India. According to information from the University of California Museum of Paleontology (UCMP), savannas are always found in warm or hot climates where the annual rainfall is from about 50.8 to 127 cm (20—50 inches) per year. Rainfall occurs within a six-to-eight-month period followed by a very long period of drought.

These tropical grasslands are home to a wide variety of animals and support a thriving ecosystem. Many medium-sized to large herbivores thrive in grasslands since there is plenty of grass to eat. There are also a lot of carnivores that roam these areas that feed on these grazing animals. Some of those that live in many well-known savannas include zebras, gazelles, wildebeests, antelopes, elephants, impalas, lions, cheetahs, and wild dogs among many others. There are also a variety of birds that roam grasslands to feed on insects that live on grass and shrubs.

The largest savannas are found in Africa. It covers more than 50% of the entire continent. Many species of flora and fauna are found here. Tourists from all over the world visit Africa to see the beautiful animals in their savannas. African safari tours allow visitors to observe animals in their natural habitat and in some cases even take photos from afar. Some lodges also allow tourists to camp out for a night or two in the middle of a grassland.

Temperate grasslands. Temperate grasslands are typically found in areas above the equator. These are wide plains that are often drier than savannas with temperatures that are a lot colder. They are flat, treeless, covered with grass, and have rich soil. This is why many temperate grasslands have been turned into farms or grazing lands for cattle.

Temperate grasslands include the Eurasian steppes in Ukraine and some parts of Russia, Argentine pampas as well as all the prairies found throughout North America and the great plains of Canada. According to UCMP, temperate grasslands have hot summers and cold winters with moderate rainfall. The temperature range in temperate grasslands is very large over the course of the year. Summer temperatures can be well over 38℃ (100 degrees Fahrenheit), while winter temperatures can be as low as -40℃ (-40 degrees Fahrenheit).

The soil in temperate grasslands are dark and covered with a very fertile and nutrient-rich layer that supports different species of grass including blue grama, purple needlegrass, and buffalo grass, as well as some flowers such as, goldenrods, blazing stars, wild indigos, coneflowers, sunflowers, clovers, and asters. The species depends on the area's amount of rainfall, temperature, and overall climate.

Numerous animals thrive in temperate grasslands according to the UCMP. These include gazelles, zebras, rhinoceroses, wild horses, lions, wolves, prairie dogs, jackrabbits, deer, mice, coyotes, foxes, skunks, badgers, blackbirds, grouses, meadowlarks, quails, sparrows, hawks, owls, snakes, grasshoppers, leafhoppers, and spiders.

2. Benefits from grasslands

SLU (The Swedish University of Agricultural Sciences), Stockholm and Lund Universities have led an international research group that has assessed ecosystem services from grasslands, using Northern Europe and Southern Africa as case studies. Here are the ten benefits from grasslands according to the assessment:

Water management. Grasslands are important for water capture and flow regulation, and decreases risks for flooding. They are also, especially in South Africa, important for the availability and accessibility of water during dry seasons.

Erosion control. Land that is permanently covered by vegetation (in contrast to most arable fields) decreases erosion and loss of topsoil. Grasslands that are not overgrazed decreases surface water runoff and stabilise the soil.

Climate. Permanent grasslands store large amount of carbon in the soil. Much more than arable land and sometimes as much as forest soils.

Fodder for animals. In South Africa as well as NW Europe grasslands are or can potentially be important as grazing lands for cattle and other animals. Grassland grazing can provide us with meat and dairy products from land that cannot be farmed.

Pollination. Pollinators like bees and bumblebees thrive in grasslands where there are many kinds of flowers. Several studies show better pollination in the vicinity of natural grasslands.

Biological pest control. Grasslands are good habitats for ladybirds, ground beetles and other beneficial insects that feed on pest insects and contribute to biological control. Studies show that biocontrol is better in landscapes containing grasslands.

Medicines. In South Africa medicinal plants from grasslands are important. More than half of the traditional medicinal plants that are traded grow in grasslands.

Tourism, recreation and hunting. Grasslands are appreciated for walking and trekking, birdwatching and hunting. They are also often popular for recreation, picnics and excursions.

Cultural history values. Ancient grasslands often contain historical legacies, like burial mounds. In Europe, grasslands remind us about our cultural history, being living remnants of the time when they were crucial for winter fodder for animals. This is still appreciated by local nature conservation societies and village communities practicing traditional management.

Biodiversity. Grasslands are valuable habitats for many plants, insects, birds and other organisms, both common and threatened species. Biodiversity is regarded as a prerequisite for many of the other ecosystem services, like pollination, biocontrol and recreation.

3. Grassland conservation

Grasslands are one of the most endangered ecosystems on Earth yet are often overlooked in discussions about conservation. Around the world, only 8% of the total grasslands that exists are protected in some way. In North America, only 1% of the grasslands are protected, even in areas that are being restored.

Grasslands are threatened by habitat loss, which can be caused by human actions, such as unsustainable agricultural practices, overgrazing, and crop clearing. Almost half of all temperate grasslands and 16% of tropical grasslands have been converted to agricultural or industrial uses and only 1% of the original tallgrass prairie exists today.

When everybody gets involved in saving the grassland biome, then more work can get done every day. Whether you speak up for wildlife or work to restore lands to their previous state, it is important to do something. We need this biome to work properly so that every species has a chance to survive and thrive. Even something as simple as not having a periodic fire is enough to threaten the very existence of this biome.

The grassland biome is one of the most expansive and diverse biomes on the planet today.

The soil is rich and fertile, but that does not mean that we should be exploiting this resource. When carefully managed, our need for farming can co-exist with our need for the biome.

Self-assessment

Read the self-assessment guidelines for *Self-assessment scale for reading comprehension* and *Self-assessment scale for written expression* (based on the *China's Standards of English Language Ability*) outlined in the appendix of this textbook. Please conduct a self-assessment to evaluate your own skills respectively.

Unit 6 Forest Park

Learning Objectives

After learning this unit, you will be able to:

- Get acquainted with the importance of forest national parks.
- Know how to deal with unfamiliar words.
- Know the basic features of an argumentation and how to write a thesis statement.

Part I Reading for Academic Purposes

Dealing with Unfamiliar Words

It is very common that when reading, you encounter vocabulary that you are not familiar with. A key point is that you don't need to interrupt your reading to look up every unfamiliar word right away in the dictionary. Mark the unfamiliar word, try the following tactics for making an "educated guess" at the meaning as you go. You'll acquire some real understanding of how the word is used.

(1) SOUND it out. When permitted, use simple phonics to attempt pronouncing the word. You might recognize the word when you hear it.

(2) BREAK it down. Look for familiar word parts, and see if you can tell how the prefixes and suffixes shape the root meaning.

(3) Identify the PART of SPEECH. In many cases, learning the part of speech and having a good idea about the function of the part of speech and learning the use of it in the syntax may be helpful.

(4) Check the CONTEXT. Guess at the word's meaning from the way it is used in the sentence. You may find that an informal definition in somewhere near. Or maybe you will see the meaning reflected in the next idea, or just be able to tell the meaning by the way the passage continues.

(5) Consult the DICTIONARY. If you can't understand what you're reading after using the above steps, pause and turn to the dictionary or the textbook's glossary list. When you find your word, skim through the whole entry and find the most relevant meaning. Check the pronunciation too.

(6) Reinforce your understanding by WRITING a usable brief definition or synonym in the margin of your reading—in pencil, because you won't always need it there.

You should also use the dictionary as a final step even if you have been able to guess well enough to keep going in your reading. When you stop after a section of reading to make notes, check your understanding of any words that aren't yet crystal-clear. Read the dictionary entry thoroughly—look for analysis of the word's derivation and structure and for examples of its usage—so that you can recognize it.

Reading Passage 1

Critical Thinking Questions

Directions: Read the following passage and answer the following questions.

1. What does clear-cutting mean?
2. Why are some people favoring the approach of clear-cutting?
3. Why are some people disapproving the clear-cutting approach?

Is Clear-cutting U.S. Forests Good for Wildlife?

Christopher Ketcham

*Critics say the idea that forests should be logged to keep them young so wildlife can thrive is based on **flawed** science.*

A. Clear-cutting in federal and state forests has long been widespread and controversial. But now, the Forest Service's Early **Successional Habitat** Creation Project represents new reasoning—which is hotly debated—that clear-cutting benefits native **fauna**. Just about every northern state east of the Great Plains, from Michigan to Maine, has implemented some version of what can be called "logging for wildlife".

B. The thinking is that clear-cutting done **judiciously** can mimic natural disturbances, for example, from insect invasions or from storms **toppling** older trees, that produce what ecologists call early successional habitat—places where young trees and shrubs **predominate** and animals that depend on such habitat thrive.

C. Timber interests are enthusiastic about the approach because it lets them profit from cutting trees while claiming the **mantle** of conservation. Hunting groups favor it because a younger, less dense forest makes it easier to find the game and birds they're tracking.

D. Porter, who is the executive director of Standing Trees, a citizen's group in Montpelier, Vermont, which lobbies for environmental protection of New England's forests, for one, is skeptical. He describes clear-cutting, which involves taking out most or all trees in a given area, as "**brutalizing** the landscape", and many other environmentalists see it the same way. "What we're doing here is going into forests that are

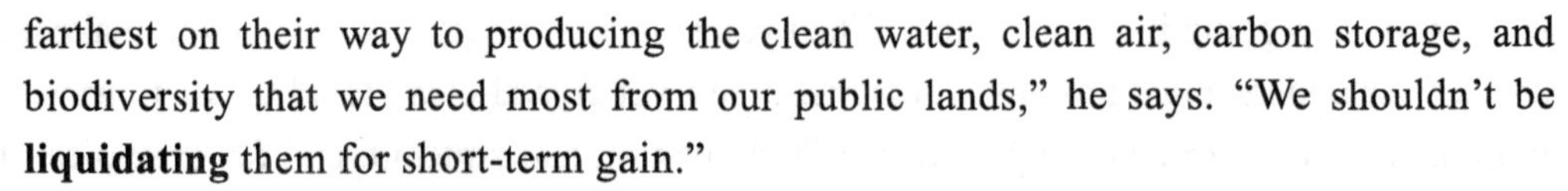

farthest on their way to producing the clean water, clean air, carbon storage, and biodiversity that we need most from our public lands," he says. "We shouldn't be **liquidating** them for short-term gain."

Young forests are for the birds

E. The point person at Mass Audubon for spreading the word about the value of early successional forest management is Tom Lautzenheiser, a 47-year-old conservation ecologist who's been with the organization since 2003. "We're in the middle of two crises, a biodiversity crisis and a climate crisis," Lautzenheiser says. "The biodiversity crisis is in part driven by habitat loss and habitat **degradation**, and one way to address that is to create habitat. That's the thinking here." He singled out the white-throated sparrow, a songbird that depends on young open woodlands whose population has been declining across the Northeast since the 1960s.

F. Yet according to the Cornell Lab of Ornithology, which collects population data on bird species in North America, all the birds identified as needing logging for wildlife—except one, the golden-winged warbler—are common, and in little danger, across their ranges. None of the game species—American woodcock, bobwhites, quails, ruffed grouse, and others—are in the least bit endangered; they're so common in fact that they're heavily hunted every year. As for the white-throated sparrow, Cornell reports that it's an "abundant" species of low conservation concern.

Young forests don't help with the climate crisis

G. In 2019, a bill, H.897, was introduced in Massachusetts that would make all state-owned public lands **off-limits** to commercial logging. It would provide sweeping forest protections no other state has ever adopted and serve as a national model to preserve wild forests intact, specifically to function as carbon sinks—places that mitigate global warming by capturing a lot of carbon.

H. That Mass Audubon joined arms with the timber industry to defeat the bill shows how **fervently** the organization believes in the logging-for-wildlife approach to managing forests. **Spearheading** opposition to H.897, Mass Audubon signed an open letter to legislators urging them not to pass it. "We do not think the best way to maximize the contribution of forests to addressing climate change is to prohibit timber harvest on all state lands," the letter said.

I. Bill Moomaw, a climate scientist and forest ecologist at Tufts University, says this is wrong-headed. Leave forests alone, Moomaw says, and carbon **uptake** will **skyrocket**. Indeed, older forests with larger trees hold **disproportionately** more carbon, according to a 2018 global study. Half of the carbon, the researchers found, is in the largest one percent of trees.

J. Undisturbed soil in old forests traps more carbon too. According to Moomaw, there's more carbon in the ground in a forest that's 150 years or older than in the standing trees themselves. That's because over the decades, annual leaf fall and blowdowns of branches transform into soil that retains ever-greater volumes of carbon.

Are young forests good for biodiversity?

K. Moomaw says the science is settled that climate **mitigation** is never served by cutting trees. But what about the claimed need for logging to "maximize biodiversity" in regions such as the Northeast? He and other scientists say this central **tenet** of the Forest Service's early successional habitat creation program also **crumbles** under critical review.

L. "Of course, a managed forest can be tailored to the needs of individual species, but that is at the expense of greater biodiversity," Moomaw says. "Greater biodiversity means not just birds but all species. It means fungi that increase the growth and carbon accumulation by forests, bacteria that build soil carbon, lichens that extract minerals from rocks, and **pollinators**, including insects."

M. Values assigned to ecosystem services and biodiversity were higher for forests defined as 170 years or older. Those forests support higher levels of carbon storage, timber growth, and species richness than younger forests. They're more resilient in the face of climate change, exhibiting what the researchers called "low climate sensitivity".

N. Other scientists are skeptical of young forest logging programs. "There's no conservation reason for creating more early successional habitat," says John Terborgh, professor **emeritus** of environmental science at Duke University, in North Carolina, and one of the world's top conservation biologists. Clear-cutting trees to expand such habitat, he said, "is a **bogus** argument, **ginned** up as an excuse for more logging."

Management for arbitrary human preference

O. Kellett believes that the adoption of the logging-for-wildlife idea is cultural landscape nostalgia turned into management practice on public lands. "This whole thing is not to restore natural habitat," he notes. "It's to maintain a human-created artificial landscape that reflects the lived memory of people alive right now. This is management for **arbitrary** human preference. The more widespread the movement becomes, the more biologists and ecologists are speaking out against it."

P. From this perspective, Zack Porter's description of logging for wildlife as **liquidation** for short-term gain—the short-term gain of favoring habitat for species people today want to see and hunt—seems apt.

Q. Just as Aldo Leopold says in the **foreword** to his *Sand County Almanac*, "We abuse land because we see it as a commodity belonging to us. When we see land as a community to which we belong, we may begin to use it with love and respect."

Key Words and Phrases

1. clear-cutting /klɪə 'kʌtɪŋ/ *n.* remove all the trees at one time 清除林木
2. flawed /flɔːd/ *adj.* blemished, damaged, or imperfect in some way 有错误的

3. successional /sək'seʃənl/	*adj.*	of or pertaining to a succession; existing in a regular order; consecutive 连续的；演替的
4. habitat /'hæbɪtæt/	*n.*	the place where a particular type of animal or plant is normally found（动植物的）生活环境，栖息地
5. fauna /'fɔːnə/	*n.*	all the animals living in an area or in a particular period of history（某地区或某时期的）动物群
6. judiciously /dʒu'dɪʃəsli/	*adv.*	carefully and sensibly; with good judgement 审慎而明智地
7. topple /'tɒp(ə)l/	*v.*	to become unsteady and fall down 倒塌，倒下
8. predominate /prɪ'dɒmɪneɪt/	*v.*	to be greater in amount or number（数量上）占优势；以……为主
9. mantle /'mæntl/	*n.*	the role and responsibilities of an important person or job 责任，职责
10. brutalize /'bruːtəlaɪz/	*v.*	to treat in a cruel or violent way 残酷对待
11. liquidate /'lɪkwɪdeɪt/	*v.*	to destroy or remove somebody or something that causes problems 清除
12. degradation /ˌdegrə'deɪʃn/	*n.*	the process of becoming worse or weaker, or being made worse or weaker 恶化；衰退
13. off-limits /ˌɒf 'lɪmɪts/	*adj.*	where people are not allowed to go 禁止进入的
14. fervently /'fɜːvəntli/	*adv.*	with very strong and sincere feelings 热诚的
15. spearhead /'spɪəhed/	*v.*	to serve as leader or leading element of 带头
16. uptake /'ʌpteɪk/	*n.*	an act or instance of absorbing and incorporating especially into a living organism, tissue, or cell 吸收
17. skyrocket /'skaɪrɒkɪt/	*v.*	suddenly increase by a very large amount 剧增；猛涨

18.	disproportionately /ˌdɪsprə'pɔːʃənətli/	*adv.*	in surprising or unreasonable amount or size when compared with something else 不成比例地
19.	mitigation /ˌmɪtɪ'geɪʃn/	*n.*	a reduction in the unpleasantness, seriousness, or painfulness 缓解
20.	tenet /'tenɪt/	*n.*	the main principle on which a theory or belief is based（理论、信仰的）基本原则
21.	crumble /'krʌmb(ə)l/	*v.*	to begin to fail or get weaker or to come to an end 崩溃
22.	pollinator /'pɒlɪneɪtə(r)/	*n.*	something that fertilizes plants with pollen, especially a type of insect 传花粉者
23.	emeritus /ɪ'merɪtəs/	*adj.*	one retired from professional life but permitted to retain as an honorary title the rank of the last office held（退休后保留的）荣誉职称
24.	bogus /'bəʊgəs/	*adj.*	pretending to be real or genuine 假的；伪造的
25.	gin /dʒɪn/	*v.*	to generate, to come up with 产生
26.	arbitrary /'ɑːbɪtrəri/	*adj.*	existing or coming about seemingly at random or by chance or as a capricious and unreasonable act of will 随意的
27.	liquidation /ˌlikwɪ'deɪʃ(ə)n/	*n.*	the action of liquidating sb./sth. 清盘；清算
28.	foreword /'fɔːwɜːd/	*n.*	a short introduction at the beginning of a book（书的）前言，序言

Reading Comprehension

Directions: Read **Passage 1** again and answer the following questions.

1. The sentence under the title and before Para. A "*Critics say the idea that forests should be logged to keep them young so wildlife can thrive is based on flawed science.*" serves as __________.

A. the thesis statement of the following passage

B. the introduction of the following passage

C. the conclusion of the following passage
D. the reference of the following passage
2. The first four paragraphs state __________.
A. the background of clear-cutting approach
B. the advantages of clear-cutting
C. the disadvantages of clear-cutting
D. the pros and cons of clear-cutting
3. In Para. H, "...joined arms with... " is closest in meaning to __________.
A. joined fight with
B. joined hands with
C. joined organizations with
D. joined army with
4. What does clear-cutting mean according to John Terborgh?
A. Clear-cutting is a reasonable approach.
B. Clear-cutting is a feasible approach.
C. Clear-cutting is a profitable approach.
D. Clear-cutting is a false approach.
5. The attitude of the writer toward clear-cutting is __________.
A. supportive
B. dissenting
C. neutral
D. not clear

Vocabulary Exercises

Exercise 1

Directions: In this section, there are ten sentences, each one with one word missing. You are required to complete these sentences with the proper form of the words given in the brackets.

1. In business, ____________ analytical thinking tends to manifest itself in a focus on growth over development, or bigger over better. (flaw)

2. Does the life evolution really only follow from simple to the complex ____________ variation process? (succession)

3. Women and girls often ____________ bear the burden of crises and conflicts. (proportion)

4. He was ____________ by the experience of being in prison. (brutal)

5. They have not hesitated in the past to ____________ their rivals. (liquid)

6. There are serious problems of land ____________ in some arid zones. (grade)

7. It's a fine line between applying too much contingency(意外) or risk ____________ and not implementing a risk response at all. (mitigate)

8. Undergraduate phase of the ____________ of many experiences for me to be a career goal of researchers has laid a good start. (accumulate)

9. Females ____________ in number over males in all the residential neighborhoods and in the suburbs of the city. (predominant)

10. A ____________ generally must have resided in the district which elects him for at least a year before election. (legislate)

Exercise 2

Directions: In this section, there are ten sentences with ten blanks. You are required to select one word for each blank from a list of choices given in a word bank. Each choice in the bank is identified by a letter. You may not use any of the words in the bank more than once.

A. habitat	B. fervently	C. extract	D. arbitrary	E. off-limits
F. judiciously	G. mitigate	H. crumble	I. resilient	J. bogus
K. skyrocket	L. spearhead	M. uptake	N. foreword	O. mantle

1. The nurse safeguards the client's right to privacy by ____________ protecting information of a confidential nature.

2. Piracy is a particular threat because of a second, bigger problem: the apparently ____________ nature of e-book pricing.

3. With these in hand, the thieves can simply set up a ____________ money transfer without tipping off (向……通风报信) the bank.

4. That may have released trapped methane (甲烷) which made global temperatures ____________ by 4—8 degrees Celsius.

5. News reports said British and French warplanes would ____________ the attack.

6. The vice-president must now take on the ____________ of supreme power.

7. With a little additional work, you can ____________ the information from the log and write that data into a database.

8. A heavy downpour caused the wall of his house to ____________.

9. Our cameras were allowed exclusive access inside the Taj Mahal (泰姬陵) normally ____________ to visitors.

10. The mechanism of nutrient ____________ by soybean roots is only part of a more general problem on the subject involving all plant species.

Exercise 3

Directions: Read the **passage 1** again and translate the following sentences from the passage (underlined in the passage) into Chinese.

1. The thinking is that clear-cutting done judiciously can mimic natural disturbances, for example, from insect invasions or from storms toppling older trees, that produce what ecologists call early successional habitat—places where young trees and shrubs predominate and animals that depend on such habitat thrive.

2. It would provide sweeping forest protections no other state has ever adopted and serve as a national model to preserve wild forests intact, specifically to function as carbon sinks—places that mitigate global warming by capturing a lot of carbon.

3. From this perspective, Zack Porter's description of logging for wildlife as

liquidation for short-term gain—the short-term gain of favoring habitat for species people today want to see and hunt—seems apt.

Part II Academic Writing Strategies

Argumentation Patterns and Writing a thesis Statement

1. Argumentation patterns

There is no set model of organization for argumentation. Table 7 are some common patterns.

Table 7 Main Patterns for Argumentation

Patterns	Description
Pattern 1	● State your first major sub-argument. ● Provide your supporting evidence. ● **Reason 1** ● **Reason 2** … ● State your second major sub-argument. ● Provide your supporting evidence. ● **Reason 1** ● **Reason 2** … ● **Refute** your opponents' first point. ● **Refute** your opponents' second point. …
Pattern 2	● **Refute** your opponents' first point. ● **Refute** your opponents' second point. … ● **Reason 1:** Provide your first major sub-argument and supporting evidence. ● **Reason 2:** Provide your second major sub-argument and supporting evidence. …
Pattern 3	● **Reason 1:** Provide your first major sub-argument and supporting evidence, which also **refutes** one of your opponents' points. ● **Reason 2:** Provide your second major sub-argument and supporting evidence, which also **refutes** one of your opponents' points. …

2. Steps for writing a thesis statement

A thesis statement should come before any detailed argumentation. After all, you can't start the writing process if you don't know what you're writing about. But just because you write it first, your thesis statement doesn't have to come out perfect right off the bat. You can take the following steps:

(1) Ask yourself a question. A good working thesis starts as a question you ask

yourself and helps guide the direction and structure of your essay. What are you trying to solve or prove in your essay? When you've identified that question, start gathering some research.

(2) Answer your question. After you've done some preliminary analysis, formulate an answer to your topic question.

(3) Develop your stance. Using more research and analysis, consider your answer again. If you're writing an argumentative essay, figure out how to convince your reader to agree with your viewpoint. Start drafting an outline to organize your points and keep your essay clear and concise.

(4) Refine your statement. Your final thesis statement needs to summarize your argument or topic. Instead of just stating your position, it should justify your point of view and set you up to make a compelling argument for it.

(5) Write your essay. Finally, it is the time to flesh out your essay. Draft an introduction and put your thesis statement at the end of the first paragraph. This gives you a chance to build up to your main point. From there, continue outlining your supporting arguments and use that as a roadmap for the rest of your essay.

Reading Passage 2

Critical Thinking Questions

Directions: Read the following passage and answer the following questions.

1. Are there any national parks in your hometown?
2. What achievements have been made in building national parks in China?
3. What are the difficulties China facing in building national parks?
4. What is your understanding of eco-civilization?

China Forges Ahead with Ambitious National Park Plan

Kyle Obermann

A. A tourism infrastructure is **popping up** among the high-elevation bamboo forests of Sichuan Province, where a new panda park is scheduled for completion by year's end. It's part of a **sprawling** new national park system of 10 pilot parks, spread across 12 provinces, whose goal is to protect habitats of endangered species. These animals range from the Siberian tiger on the Russian border to the world's last 30 Hainan black **crested gibbons** in southern China's tropical rainforest. By uniting hundreds of protected areas managed by various municipalities and provinces, the new system's goal is to **streamline** and strengthen conservation under the central authority of the new National Forestry and Grassland Administration. Already, these test parks—which should be officially approved by the end of 2020—cover an area two-thirds the size of the U.S. national park system. The largest, Sanjiangyuan National Park in Qinghai Province, is about the size of Mississippi. In 2019, when China held its first national park conference, Xi Jinping issued a rare public letter of support for the project.

B. With such an ambitious plan, there are bound to be **hurdles**—and among the most **formidable** are working with local people and balancing the need for tourism with wildlife conservation. For instance, China has only offered voluntary relocation to a **fraction** of the 652,600 people living inside the 10 parks, hoping instead that existing communities will welcome **ecotourism** and embrace the new network of protected areas, which is modeled in part on the U.S.'s national park system. For this vision to succeed, so must the people who rely on the land, says Li Xinrui, who helps manage a community cooperative within the Guanba Community Nature Reserve. "Whether or not your protection efforts are effective doesn't depend upon how well a nature reserve or national park is created, but on whether the livelihoods of the locals have changed," Li says. "When the common people can have good income and lives, that's when protection efforts will be effective in conserving nature."

Lone tourist

C. Already, many people living in existing Chinese nature reserves work in ecotourism, a $3 billion industry that serves 128 million visitors a year. But there's still a long way to go. For one, ecotourism—which is defined as tourism that benefits both locals and their environment—only exists in a fifth of China's nature reserves. What's more, the government has not announced a plan to create a **backcountry** camping and permitting system, which would regulate how people enjoy nature. Jennifer Turner, director of the China Environment Forum at the Wilson Center, a **think tank** based in Washington, D.C., adds, "there are not even immediate plans to have national park **rangers**." Some local governments are hiring rangers, but there is no formal structure and training across the different pilot projects, she notes.

Doing tourism right

D. Yet there's hope that existing ecotourism projects will inspire efforts elsewhere in the new park system, experts say. Take the remote Tibetan village Angsai, which lies on the banks of the upper Mekong River within Sanjiangyuan National Park. Since 2018, the village has run a community-led tourism program that benefits both locals and snow **leopards**, the region's main tourist draw. For $43 a day, visitors can stay with local Tibetan families, who act as guides to spot these rare "ghost cats" in the wild. About 75 percent of visitors who stay at least three days see the **feline**, according to Terry Townshend, a consultant with Shan Shui Conservation Center, one of China's largest conservation nonprofits, and an advisor for the Paulson Institute. "The community has made all the major decisions, and 100 percent of the revenue stays in the community," Townshend says. "It's been incredibly successful." In 2019, Angsai became the first community tourism **franchise** in a pilot park to be approved by the federal government, he adds. "It was **showcased** as a way of doing tourism in environmentally sensitive places."

E. Marc Brody, a National Geographic Explorer and founder of Panda Mountain, an eco-tourism and conservation organization, who has worked in China since 1994, agrees that well-designed ecotourism, such as in Angsai and in Wolong Nature Reserve, home to the Wolong Panda Center, can boost local ecosystems. "A core mission of China's nation- al parks is to promote eco-civilization—a mission that can be advanced by involving visitors," says Brody. Written into the national constitution in 2012, eco-civilization means sustainably balancing the economy and the environment. "The process of engaging people in habitat **restoration** is a way for people to see the landscape in a more holistic and **interdependent** manner," he says, "and restore hope that we can help save endangered species."

Key Words and Phrases

1. pop up		appear in a place or situation suddenly or unexpectedly 突然出现
2. sprawling /sprɔːlɪŋ/	*adj.*	spreading out carelessly (as if wandering) in different directions 蔓延的；杂乱无序伸展的
3. crested gibbon		冠长臂猿
4. streamline /'striːmlaɪn/	*v.*	to make an organization or process more efficient by removing unnecessary parts 提高效率
5. hurdle /'hɜːdəl/	*n.*	a problem, difficulty, or part of a process that may prevent you from achieving something 障碍；困难
6. formidable /fə'mɪdəb(ə)l/	*adj.*	extremely impressive in strength or excellence, inspiring fear 可怕的；难对付的
7. relocation /ˌriːləʊ'keɪʃn/	*n.*	the transportation of people (as a family or colony) to a new settlement (as after an upheaval of some kind) 重新安置
8. fraction /'frækʃn/	*n.*	a small part or amount 小部分；少量；一点儿
9. ecotourism /'iːkəʊtʊərɪzəm/	*n.*	the practice of touring natural habitats in a manner meant to minimize ecological impact 生态旅游
10. backcountry /'bækkʌntri/	*n.*	a remote undeveloped rural area 偏僻地区，偏远地区

11.	think tank		a group of experts who provide advice and ideas on political, social or economic issues 智囊团，专家组
12.	ranger /'reɪndʒə(r)/	*n.*	a person whose job is to take care of a park, a forest or an area of countryside 园林管理员；护林人
13.	leopard /'lepəd/	*n.*	a large animal of the cat family, that has yellowish-brown fur with black spots 豹
14.	feline /'fiːlaɪn/	*n.*	an animal of the cat family 猫科动物
15.	franchise /'fræntʃaɪz/	*n.*	a business established or operated under an authorization to sell or distribute a company's goods or services in a particular area 获特许权的商业（或服务）机构
16.	showcase /'ʃəʊkeɪs/	*v.*	to exhibit especially in an attractive or favorable aspect 展示
17.	restoration /ˌrestə'reɪʃ(ə)n/	*n.*	the act of restoring something or someone to a satisfactory state 恢复
18.	interdependent /ˌɪntədɪ'pendənt/	*adj.*	that depend on each other; consisting of parts that depend on each other （各部分）相互依存的

Writing Tasks

Exercise 1

Directions: Read the following sentences from **Passage 2** and rewrite the following sentences(underlined in the passage), replacing the underlined parts with your own words.

1. A tourism infrastructure is popping up among the high-elevation bamboo forests of Sichuan Province, where a new panda park is scheduled for completion by year's end.

2. With such an ambitious plan, there are bound to be hurdles—and among the most formidable are working with local people and balancing the need for tourism with wildlife conservation.

3. It was showcased as a way of doing tourism in environmentally sensitive places.

4. The process of engaging people in habitat restoration is a way for people to see the landscape in a more holistic and interdependent manner and restore hope that we can help save endangered species.

Exercise 2

Directions: Identify the topic sentence of each paragraph in **Passage 2** and write them down.

Para. A ____________________

Para. B ____________________

Para. C ____________________

Para. D ____________________

Para. E ____________________

Exercise 3

Directions: Read **Passage 2** again and sum up the main idea within one sentence as the thesis *statement.*

____

Exercise 4

Directions: Read the passage again and write a summary of **Passage 2** with no more than 150 words.

Extensive Reading and Writing

Directions: For this part, you are required to read the following passage and then write an argumentation on **Building Urban Forest Parks: A Feasible Way out of the Climate Crisis**. You should write at least 120 words but no more than 180 words.

As the Climate Crisis Worsens, Cities Turn to Parks

Sarah Gibbens

Cities across the U.S. are seeing worse floods and hotter summers, but experts believe urban parks can help residents cope.

City parks have long been a place for urban residents surrounded by the gray of asphalt and concrete to get a small dose of green. As cities increasingly feel the impacts of rising seas and temperatures, city planners are rethinking the roles of urban parks.

"There's been a quiet and profound move to use parks to help cities adapt to the realities of climate change," says Diane Regas, CEO of The Trust for Public Land, an organization that works to create neighborhood and national parks.

Each year the trust publishes their ParkScore Index, which ranks the top 100 largest U.S. cities by parks. The 2019 rankings will be released on Wednesday; Minneapolis won the highest ranking in 2018. The Trust looks at size, convenience, amenities, and financial investment to compile its list.

While amenities like basketball hoops and playgrounds have long been assets that bumped cities into top spots, increasingly, Regas says, The Trust is seeing cities build parks that can alleviate climate change effects like intense heat, flooding, and poor air quality.

And it's more than shade trees that are helping fight climate change. The Trust says parks can help mitigate coastal flooding, capture carbon, and foster a sense of community among those that will be affected by extreme weather.

Cooling down islands of heat

All of the dark-gray asphalt in cities collect heat—a lot of it. A 2018 study by the National Oceanic and Atmospheric Administration mapped the hottest areas of Washington, D.C. and found that intense heat nearly always aligned with the densest urban areas. Large parks cooled certain parts of the city by as much as 17 degrees Fahrenheit. That kind of cooling can be a lifesaver given the more than 600 annual deaths caused by heat-related illnesses.

Dallas is one of the country's fastest-warming cities thanks in part to its sprawling, impervious surfaces, but with a new $312 million bond package, the city is hoping to change that.

"A lot of people see Texas as very conservative but there's no denying that climate change is real, and our cities understand that," says Dallas Park and Recreation President Robert Abtahi. "Our parks are taking on that big challenge in the ways we can the best."

Using satellite data, the city is able to see what neighborhoods most need the cooling effect of green spaces. Parks like the planned Dallas Water Gardens will be situated in some of the most heat-stressed parts of the city.

Mopping up floods

Cities are increasingly being inundated with floodwaters, and city planners think parks can help with this issue too.

A report published in February by *The Nature Conservancy* looked at the best ways to mitigate flooding in Houston, a city with many neighborhoods built on floodplains and regularly inundated by rising waters. Offering affected homeowners buyouts and converting homes into green spaces would save more money than installing infrastructure like pipes, they found. The coastal city has projects under way to create more than 150 miles of trails, parks, and green spaces along the many bayous running through its urban landscape. While Houston received a wakeup call from Hurricane Harvey in 2017, it was Hurricane Sandy's 2012 strike along the Northeast coast that forced cities to reconcile their new realities.

New York City is building a patchwork of waterfront parks designed to absorb the energy of incoming water and to drain deluges of rainwater. Atlanta, after repeatedly being hit by flash floods, is creating a 16-acre park designed to absorb millions of gallons of water.

In Boston, Parks and Recreation Commissioner Christopher Cook says the city is preparing to see 40 inches of sea level rise by 2050.

"Parks are playing an outsized role in the adaptation plan," he says, but emphasizes that the city doesn't see urban parks as a solution to stop or reverse climate change. For that, he touts instead Boston's plan to become carbon neutral over the next 30 years.

One piece of the puzzle

Brendan Shane, the climate program director at the Trust for Public Land, says parks can ultimately provide a sort of social resilience, in addition to cooling neighborhoods and absorbing floodwater.

"The stronger the bonds are from neighbor to neighbor, the better they are able to react to a shock," he says. "The nice thing about parks is they give you all those things at the same time."

While drainage pipes and reservoirs have also been used to curb some of the impacts of rising seas, the Trust and city planners see parks as a way to adapt while providing a better quality of life. It's not only about making green space, they say, but also about creating opportunities for people to exercise and play.

"Not a single solution by itself will avoid the climate crisis. We see parks as an important part of it, but there isn't a silver bullet," says Regas. She adds: "Parks are an example of what we in the environmental community need to do to embrace solutions that simultaneously address climate change and make people's lives better."

Self-assessment

Read the self-assessment guidelines for *Self-assessment scale for reading comprehension* and *Self-assessment scale for written expression* (based on the *China's Standards of English Language Ability*) outlined in the appendix of this textbook. Please conduct a self-assessment to evaluate your own skills respectively.

Unit 7 Forestry Information Technology

Learning Objectives

After learning this unit, you will be able to:

- Know about the forestry information technology.
- Improve your understanding of long sentences.
- Learn how to support your claim.

Part I Reading for Academic Purposes

Reading Long Sentences

We are often faced with the difficulties in understanding complicated sentences while reading academic material. However, this is an obstacle that we readers must overcome if we want to touch the core of science and technology. Successful readers can interpret a writer's message by recognizing and analyzing the syntax of sentences, which means, the sentence structure. Generally speaking, a sentence is composed of a "main body": the subject, predicate and object, with various supplementary parts added to this main body to supply further information. For example, in this sentence:

> Whereas decades ago, this management would be tough, effort-demanding work, nowadays, geographic information systems (GIS) facilitate this work as they allow capturing, storing, quantifying and transforming data into useful information that can be edited and displayed for specific requirements or decisions by local managers.

The main body is "geographic information systems facilitate this work". The added ideas are "as they allow capturing, storing, quantifying and transforming data into useful information". Furthermore, the information "can be edited and displayed by local managers".

Hence, we understand that it is essential to look for the "main body" of the sentence to understand the message. Only by acquiring the primary idea of the sentence can we analyze what parts are added to this basic information. We should also get familiar with certain grammar patterns of written structures such as tenses, aspects, voices, subordination, coordination, and apposition. Other syntactic devices, i.e. anaphora and placement of modifiers, can be further included in our reading experience. By reading extensively and thinking critically through comprehension of complex sentences, we gradually accumulate the competence to appreciate the richness of academic papers.

Reading Passage 1

 **Critical Thinking Questions**

Directions: Read the following passage and answer the following questions.

1. What does "GIS" mean? Why do the authors bring "GIS" into discussion?
2. What benefits does GIS bring to us?
3. What makes GIS a powerful tool to assist park management?
4. What is the right message about NbS to climate change according to the authors?

A GIS Supported Multidisciplinary Database for the Management of UNESCO Global Geoparks

Daniel Ballesteros, Pablo Caldevilla, et al.

A. The governance of a UNESCO Global Geopark (UGGp) is a major challenge mainly due to the participation of many unequal factors (local people, managers, regional to local public administrations, businesspersons, cultural associations and scientists) and the mixing of potential conflicting issues involved with the suitable development, such as economic factors and the conservation of nature. The progress in scientific knowledge is the basis for growth **sustained** in time, as well as other issues, including tourism marketing. However, the management of scientific data is complex due to the great scope and depth of **thematic** areas involved (e.g., geography, geology, biology, soil sciences, archaeology, history, **ethnography**, mining and engineering) and the varied **spatiotemporal** scales of different studies. Whereas decades ago, this management would be tough, effortdemanding work, nowadays, geographic information systems (GIS) **facilitate** this work as they allow capturing, storing, quantifying and transforming data into useful information that can be edited and displayed for specific requirements or decisions by local managers.

B. GIS allows the input of data from external sources such as the GIS datasets provided by European governments, following the INSPIRE European Council Directive 2007/2/CE, or from its acquisition by experts through fieldwork, remote sensing and other ways (e.g., Google Street View). Databases in GIS are organised by combining **vector** and **matrix** files (named all of their **coverages**), each one with specific information. Vector coverages (points, **polylines** and **polygons**) combine spatial and **attribute** data. In each coverage, the spatial data describes the absolute location of each geometric feature (e.g., **geosite**, tourism office), including its shape, size, coordinates and orientation. The attribute table describes details of the spatial feature (e.g., lithology, age), which are stored in tables in which each row is a feature and each column an attribute of the feature. **Raster** coverages contain a numerical value per **pixel** and are often used for representing **terrain** properties (e.g., **altitude**, **slope** and **reflectivity**). GIS offers the possibility to quantify each feature using measurement tools and statistical analyses, recognising changes in an area over a period of time (e.g., land use changes).

C. Besides information storage (that can be accessed remotely), what makes GIS a powerful tool is its analysis system, which allows extracting the required information, grouping it according to defined criteria, discarding what is not useful and transforming data into new coverages by mathematical operations. From raw data, thematic maps can be drawn, which is the end product for different users, containing all the relevant information in a simplified and user-friendly way.

D. Despite the advantages of GIS for spatial data management, its **implementation** in UGGp management is still underused and is mainly applied to the **elaboration** of geological, geosites and touristic maps, as well as for geoheritage and geodiversity studies and other research in geoparks. In a few cases, land use practices were integrated into the UGGp management following the analysis of land use changes and natural risks with GIS, as performed in the Batur volcanic UGGp in Indonesia. **Sporadically**, geoparks have used GIS for other technical purposes, such as the Sardinian Mining UGGp in Italy, where GIS was involved in a regional **delineation** and development model, the Hondsrug UGGp in the Netherlands, where the spatial **affinity** of inhabitants within the UGGp was analysed via GIS, or the Ciletuh-Palabuhanratu UGGp in Indonesia, which performed a visual analysis in GIS. Furthermore, a small number of UGGps have started the development of a tourism geographic information system (TGIS) for travel information inquiry, making specific tourism charts, and helping take decisions on tourism development, as in the case of Mount Sanqingshan UGGp in China. In recent years, selected coverages are transferred to visitors through mobile applications and web viewers such as WebGIS and Google Maps, facilitating access to information for tourists.

E. This paper presents a **pragmatic** GIS database to assist in the suitable governance of the Courel Mountains UGGp in Northwest Spain. The database is structured in 66 coverages compiled from public sources and previous works or produced through traditional mapping (combining fieldwork and photointerpretation) and GIS tools. The acquired data was later **homogenized** and **validated** by a multidisciplinary team and **archived** in independent coverages. Forty thematic maps illustrate the broad range of cartographic information included in the GIS database. Among them, 25 basic maps provide an overview of the UGGp and 15 new maps focus on crosscutting and technical issues. All maps illustrate the huge potential of GIS to create new resources combining coverages and adapting the legend according to their purpose and audience. The database facilitates the suitable publishing of consistent outputs (e.g., brochures, books, panels, webpages, web serves), as well as the elaboration of technical data to assist the park management. The database furnishes information on the design of education actions, touristic routes, activities and Geopark facilities. The GIS database is also a supportive tool for scientific research and provides the necessary knowledge to conduct geoconservation actions based on land use, geological hazards and the occurrence of natural and cultural heritages. Altogether, the GIS database constitutes a powerful instrument for policy-making, facilitating the identification and evaluation of alternative strategy plans.

Key Words and Phrases

1. sustain /səs'teɪn/ — *v.* cause to continue or be prolonged for an extended period or without interruption 保持；使持续不断
2. thematic /θɪ'mætɪk/ — *adj.* having or relating to subjects or a particular subject 主题的；专题的
3. ethnography /eθ'nɒgrəfi/ — *n.* the scientific description of peoples and cultures with their customs, habits, and mutual differences 人种志；民族志
4. spatiotemporal /speɪʃɪəʊ'tempərəl/ — *adj.* belonging to both space and time or to space-time 时空的
5. facilitate /fə'sɪlɪteɪt/ — *v.* make (an action or process) easy or easier 使（行动，过程）更容易
6. vector /'vektə(r)/ — *n.* a quantity having direction as well as magnitude, especially as determining the position of one point in space relative to another 矢量
7. matrix /'ɪneɪtrɪks/ — *n.* an arrangement of numbers, letter, or signs in rows and columns that you consider to be one amount, and that you use in solving mathematical problems 矩阵
8. coverage /'kʌvərɪdʒ/ — *n.* the extent to which something deals with or applies to something else 涉及面；探讨范围
9. polyline /'pɒlɪlaɪn/ — *n.* a continuous line that is composed of one or more connected straight line segments, which, together, make up a shape 多线段；折线

10. polygon /'pɒlɪgən/ *n.* a flat shape with three or more sides 多边形

11. attribute /ə'trɪbju:t/ *n.* a quality or feature, especially one that is considered to be good or useful 特性，属性

12. geosite /'dʒi: əʊsaɪt/ *n.* a place of geological interest 地质遗迹

13. raster /'ræstə(r)/ *n.* a rectangular pattern of parallel scanning lines followed by the electron beam on a television screen or computer monitor 光栅

14. pixel /'pɪks(ə)l/ *n.* a minute area of illumination on a display screen, one of many from which an image is composed of 像素

15. terrain /tə'reɪn/ *n.* a stretch of land, especially with regard to its physical features 地面，地形，地势

16. altitude /'æltɪtju:d/ *n.* the height of an object or point in relation to sea level or ground level 高度；海拔

17. slope /'sləʊp/ *n.* a difference in level or sideways position between the two ends or sides of a thing 斜度；坡度

18. reflectivity /ˌri:flek'tɪvətɪ/ *n.* the property of reflecting light or radiation, especially reflectance as measured independently of the thickness of a material 反射率，反射性

19. implementation /ˌɪmpləmen'teɪʃən/ *n.* the process of making something active or effective 贯彻；实行

20. elaboration /ɪˌlæbə'reɪʃən/ *n.* the process of developing or presenting a theory, policy, or system in further details 详尽阐述

21. sporadically /spə'rædɪklɪ/ *adv.* occasionally or at irregular intervals 零星地

22. delineation /dɪˌlɪni'eɪʃən/ *n.* the action of describing or portraying something precisely 准确描述

23. affinity /ə'fɪnɪti/ *n.* a similarity of characteristics suggesting a relationship, especially a resemblance in structure between animals, plants, or languages 类同，相似性

24. pragmatic /præg'mætɪk/ *adj.* dealing with things sensibly and realistically in a way that is based on practical

rather than theoretical considerations 讲究实际的，实干的

25. homogenize /hɒ'mɒdʒənaɪz/ *v.* to bring all data into a common geo-spatial framework to ensure consistency of data, integrity of analysis and validity of results 同质化

26. validate /'vælɪdeɪt/ *v.* check or prove the validity or accuracy of something 验证

27. archive /'ɑːkaɪv/ *v.* place or store in such a collection or place 存入档案

Reading Comprehension

Directions: Read **Passage 1** again and answer the following questions.

1. Why is GIS useful according to Para. B?

A. To get scientific data.

B. To edit and display data.

C. To support the management of data.

D. To provide and input data.

2. What is the main idea of Para. C?

A. It describes how GIS are organized.

B. It describes different types of data.

C. It describes how data is measured.

D. It introduces features of different data.

3. Which is NOT the advantage of GIS according to Para. D?

A. It can transform data into new coverages.

B. It can store data and access data remotely.

C. It can simplify data for different users.

D. It can draw thematic maps from raw data.

Vocabulary Exercises

Exercise 1

Directions: In this section, there are ten sentences, each one with one word missing. You are required to complete these sentences with the proper form of the words given in the brackets.

1. The oil which is ____________ from olives is used for cooking. (extract)

2. Consultants found the experience frustrating—their reports were only partly ____________, or, worse still, just pigeonholed. (implementation)

3. In Greek mythology, Narcissus fell in love with his own ____________ in a pool of water. (reflectivity)

4. The project focuses on the ____________ of regional trade and investment. (facilitate)

5. ____________ studies and the analysis of life histories of older people have enabled the mapping of family networks and their exchanges over time. (ethnography)

6. A large international meeting was held with the aim of promoting ____________ development in all countries. (sustain)

7. ____________ waste must be properly disposed of. (hazard)

8. TV has ____________ the culture and language of large parts of the planet. (homogenize)

9. The boundary of the park is ____________ by a row of trees. (delineation)

10. More than 100 people have been killed this year in ____________ outbursts of ethnic violence. (sporadically)

Exercise 2

Directions: In this section, there are ten sentences with ten blanks. You are required to select one word for each blank from a list of choices given in a word bank. Each choice in the bank is identified by a letter. You may not use any of the words in the bank more than once.

A. sporadically	B. pragmatic	C. affinity	D. archived
E. validate	F. attribute	G. attitude	H. altitude
I. matrix	J. thematic	K. geosite	L. elaboration
M. terrain	N. homogenised	O. spatiotemporal	

1. We had to drive over some rough ____________.

2. This spot offers 36 types of craft beers, including ____________ seasonal offerings.

3. To add creditability, firms should engage one of the numerous third-party firms qualified to ____________ such results.

4. She ____________ her e-mail messages in a folder on her hard drive.

5. His photographs require no written ____________—everything is in the images themselves.

6. Some visitors find it difficult to adjust to the city's high ____________.

7. Organizational ability is an essential ____________ for a good manager.

8. Europe is remaking itself politically within the ____________ of the European Community.

9. In business, the ____________ approach to problems is often more successful than an idealistic one.

10. She has a(n) ____________ to him because of their common musical interests.

Exercise 3

Directions: Read the passage again and translate the following sentences from **Passage 1** (underlined in the passage) into Chinese.

1. The database is structured in 66 coverages compiled from public sources and previous works or produced through traditional mapping (combining fieldwork and photo interpretation) and GIS tools.

2. All maps illustrate the huge potential of GIS to create new resources combining coverages and adapting the legend according to their purpose and audience.

3. The GIS database is also a supportive tool for scientific research and provides the necessary knowledge to conduct geoconservation actions based on land use, geological hazards and the occurrence of natural and cultural heritages.

Part II Academic Writing Strategies

How to support your claim

In academic writing, we aim to spread ideas so we often make claims and argue for our claims. An argument includes three parts (SES) in its simplest form: a claim or a major statement, elaboration, and specifics. Take the following paragraph as an example:

> Second, remotely sensed data are available everywhere and often at a range of spatial and temporal scales. Key environmental remote-sensing systems, such as those carried by the Landsat satellites, have provided a constantly updateable stream of imagery for the entire planet since the 1970s. Availability can sometimes be constrained by technical problems or cloud cover, but in principle, imagery should be available everywhere irrespectively of location, enabling inter alia study of sites no matter how remote or hazardous they might be. Furthermore, historical remote-sensing data allow us to go back in time to look at the causes of present environmental issues.

This claim or major statement is "remotely sensed data are available everywhere and often at a range of spatial and temporal scales". It is inadequate to simply give a sub-claim; we need to offer evidence to support this sub-claim. The evidence is following: "remote-sensing systems...have provided a constantly updateable stream of imagery for the entire planet since the 1970s". Obviously, this is an objective fact, not the author's opinion or standing.

Evidence gives claims proof or validation. It can be in the form of research data, quotes, or textual evidence from a piece of literature. It should not be a guess, assumption, or assertion based on the writer's opinion. It should mention the source from where the evidence was obtained through a citation.

Reading Passage 2

Critical Thinking Questions

Directions: Read the following passage and answer the following questions.

1. What is the writing style of the following passage?
2. Why does the author use "ubiquity" to describe remote sensing technology?
3. How does the writer support his/her claim?

Applications in Remote Sensing to Forest Ecology and Management

Alex M. Lechner, Giles M. Foody, Doreen S. Boyd

Introduction

A. Remote sensing is the acquisition of information about some feature of interest without coming into direct contact with it. Popular forms of remote sensing used in the environmental sciences are images of the Earth's surface acquired from sensors **mounted** on airborne and spaceborne platforms. Remote sensing has been used for mapping the **distribution** of forest ecosystems, global **fluctuations** in plant productivity with season, and the three-dimensional (3D) structure of forests.

B. The range and diversity of sensing systems, as well as the variety of applications, have evolved greatly over the last century. The types of images used range widely from conventional **aerial** photographs that capture a view similar to that observed by the human eye to images that reveal elements that might be invisible to the human eye, such as the physical structure and chemical composition of the Earth's surface.

C. Remotely sensed **imagery** provides a view of the Earth's surface in such a way that allows features on it to be identified, located, and characterized. Moreover, although each image provides a snapshot of the environment, it is commonly possible to acquire imagery repeatedly in time. As a result, remote sensing has been used in a diverse range of forest ecology and management applications from mapping **invasive** species to monitoring land-cover changes, such as habitat **fragmentation**, to estimating biophysical and biochemical properties of forests.

D. Given that many forest environmental variables can be estimated directly in the field, why has remote sensing become an important data source? We note six key reasons for this situation.

The Ubiquity of Remote Sensing: Six Key Reasons

E. First, remotely sensed imagery provides a **synoptic** view. The **vantage** provided by an Earth-observing sensor ensures that imagery captures a complete picture of the environment in its field of view. Thus, every visible feature, including its location and its location relative to that of all others in the **imaged** area, is captured. In short, this gives imagery a map-like format that provides a complete survey of the imaged area rather than field data, which are often based on a very limited set of samples from which inter-sample site information would have to be **inferred** by some form of **interpolation**. Because of this complete survey, remote sensing allows wall-to-wall mapping and monitoring of important ecological variables, such as land-cover change.

F. Second, remotely sensed data are available everywhere and often at a range of spatial and temporal scales. Key environmental remote-sensing systems, such as those carried by the Landsat satellites, have provided a constantly updateable stream of imagery for the entire planet since the 1970s. <u>Availability can sometimes be constrained by technical problems or cloud cover, but in principle, imagery should be available everywhere **irrespectively** of location, enabling ***inter alia*** study of sites no matter how remote or hazardous they might be.</u> Furthermore, historical remote-sensing data allow us to go back in time to look at the causes of present environmental issues.

G. Third, remotely sensed imagery has a high degree of **homogeneity**. Critically, data from key environmental remote-sensing systems are acquired under relatively fixed conditions, and the data captured relate to the way in which radiation interacts with the environment, which is constant in space and time; there are no human-induced complications, such as differences in measurement practices from one country to another.

H. Fourth, the imagery contains, or can easily be **converted** to, digital images and as such can be easily integrated with other spatial datasets in a geographical information system.

I. Fifth, per unit area, remote sensing is an inexpensive way to acquire data. Although the financial costs associated with remote sensing can sometimes be very large—for example, it is expensive to build, launch, and operate satellite remote-sensing systems, making some imagery expensive—much is freely available. Additionally, although commercial remote-sensing systems can appear costly, the data still provide inexpensive assessment on a unit area basis. More critically, however, there has been an increasing trend to make key data-sets for environmental science research freely and openly available. For example, the complete archive of the influential Landsat series of satellites is freely available, and recently the European Space Agency (ESA) launched a **suite** of new satellites and provides the data collected for free. Resources such as Google Earth Engine (GEE) also provide easy access to vast global datasets.

J. Sixth and finally, not only are data more readily available, but there has also been an increasing trend toward the provision of data products as well as the image data themselves. This reduces both the need for expert knowledge of remote sensing and image analysis and the communication gap between experts and environmental scientists, which has historically been a concern. Environmental scientists can now easily access science quality data products obtained from remote sensing (e.g., leaf area index, land use, and land cover), although expert knowledge might still sometimes be needed.

Key Words and Phrases

1. mount /'maʊnt/ *v.* to place or fix (an object) on an elevated support 安装于高处
2. distribution /ˌdɪstrɪ'bju:ʃən/ *n.* the way in which something is shared out among a group or spread over an area 分布
3. fluctuation /ˌflʌktjʊ'eɪʃən/ *n.* rise and all irregularly in number or amount 不规则波动，起伏
4. aerial /'eəriəl/ *adj.* existing, happening, or operating in the air 在空中的
5. imagery /'ɪmɪdʒəri/ *n.* pictures produced by an imaging system 图像
6. invasive /ɪn'veɪsɪv/ *adj.* (especially of plants or a disease) tending to spread prolifically and undesirably or harmfully（尤指植物或疾病）入侵的，扩散的
7. fragmentation /ˌfrægmen'teɪʃən/ *n.* the process or state of breaking or being broken into small or separate parts 破裂，碎裂
8. synoptic /sɪ'nɒptɪk/ *adj.* affording a general view of a whole 纵观全局的
9. vantage /'vɑ:ntɪdʒ/ *n.* a position giving a strategic advantage 优势
10. image /'ɪmɪdʒ/ *v.* make a visual representation of something by scanning it with a detector or electromagnetic beam（以探测器或电磁束）扫描出直观图
11. infer /ɪn'fɜ:(r)/ *v.* deduce or conclude from evidence and reasoning rather than from explicit statements 推断，推论
12. interpolation /ɪnˌtɜ:pə'leɪʃn/ *n.* the process of calculating an approximate value based on values that are already known 估算
13. irrespectively /ɪrɪ'spektɪvli/ *adv.* not taking something into account, regardless of 不考虑
14. *inter alia* among other things 除了其他事物以外
15. homogeneity /ˌhɒmədʒə'ni:əti/ *n.* the quality or state of being of a similar kind or of having a uniform structure or composition throughout 同质性
16. convert /kən'vɜ:t/ *v.* to change from one form or function to another 转变，转化
17. suite /swi:t/ *n.* a group of things forming a unit or constituting a collection 一组

Writing Tasks

Exercise 1

Directions: Read the following sentences from **Passage 2** and rewrite the following sentences (underlined in the passage), replacing the underlined parts with your own words.

1. The types of images used range widely from conventional aerial photographs that capture a view similar to that observed by the human eye to images that reveal elements that might be invisible to the human eye, such as the physical structure and chemical composition of the Earth's surface.

2. As a result, remote sensing has been used in a diverse range of forest ecology and management applications from mapping invasive species to monitoring land-cover changes, such as habitat fragmentation, to estimating biophysical and biochemical properties of forests.

3. Availability can sometimes be constrained by technical problems or cloud cover, but in principle, imagery should be available everywhere irrespectively of location, enabling *inter alia* study of sites no matter how remote or hazardous they might be.

Exercise 2

Directions: Read **Passage 2** again and answer the questions below. Please give brief answers in about 10 words.

1. What is remote sensing? (Para. A)
2. Why is remote sensing useful in monitoring land-cover changes? (Para. C)
3. Is remote sensing expensive? (Para. J)
4. Why can we look back to investigate causes of present environmental issues? (Para. G)

Exercise 3

Directions: Read the passage again and write a summary of **Passage 2** with no more than 150 words.

Extensive Reading and Writing

Directions: For this part, you are required to read the following passage and then write an essay on **Pros and Cons of GIS**. You should write at least 120 words but no more than 180 words.

Benefits and Drawbacks of GIS

Fuse GIS

GIS benefits organizations of all sizes and in almost every industry. There is a growing interest in and awareness of the economic and strategic value of GIS, in part because of more standards-based technology and greater awareness of the benefits demonstrated by GIS users. The number of GIS enterprise solutions and IT strategies that include GIS are growing rapidly. The benefits of GIS generally fall into five basic categories:

Cost savings resulting from greater efficiency

These are associated either with carrying out the mission (i.e., labor savings from automating or improving a workflow) or improvements in the mission itself. A good case for both of these is Sears, which implemented GIS in its logistics operations and has seen dramatic improvements. Sears considerably reduced the time it takes for dispatchers to create routes for their home delivery trucks (by about 75%). It also benefited enormously in reducing the costs of carrying out the mission (i.e., 12%—15% less drive time by optimizing routes). Sears also improved customer service, reduced the number of return visits to the same site, and scheduled appointments more efficiently.

Better decision making

This typically has to do with making better decisions about location. Common examples include real estate site selection, route/corridor selection, zoning, planning, conservation, natural resource extraction, etc. People are beginning to realize that making the correct decision about a location is strategic to the success of an organization.

Improved communication

GIS-based maps and visualizations greatly assist in understanding situations and story telling. They are a new language that improves communication between different teams, departments, disciplines, professional fields, organizations, and the public.

Better geographic information recordkeeping

Many organizations have a primary responsibility of maintaining authoritative records about the status and change of geography (geographic accounting). Cultural geography examples are zoning, population census, land ownership, and administrative boundaries. Physical geography examples include forest inventories, biological inventories, environmental measurements, water flows, and a whole host of geographic accountings. GIS provides a strong framework for managing these types of systems with full transaction support and reporting tools. These systems are conceptually similar to other information systems in that they deal with data management and transactions, as well as standardized reporting (e.g., maps) of changing information. However, they are fundamentally different because of the unique data models and hundreds of specialized tools used in supporting GIS applications and workflows.

Managing geographically

In government and many large corporations, GIS is becoming essential to understand what is going on. Senior administrators and executives at the highest levels of government use GIS information products to communicate. These products provide a visual framework for conceptualizing, understanding, and prescribing action. Examples include briefings about various geographic patterns and relationships including land use, crime, the environment, and defense/security situations. GIS is increasingly being implemented as enterprise information systems. This goes far beyond simply spatially enabling business tables in a DBMS. Geography is emerging as a new way to organize and manage organizations. Just like enterprise-wide financial systems transformed the way organizations were managed in the 60's, 70's, and 80's, GIS is transforming the way that organizations manage their assets, serve their customers/citizens, make decisions, and communicate. Examples in the private sector include most utilities, forestry and oil companies, and most commercial/retail businesses. Their assets and resources are now being maintained as an enterprise information system to support day-to-day work management tasks and provide a broader context for assets and resource management.

However, the drawbacks of GIS are also obvious, mainly reflected in the following aspects.

Prohibitive cost

Smaller businesses and small government offices tend to think that they can't afford GIS. Instead of software, they resort to using PDFs and manually mapping data. The software itself can be expensive, but implementing a GIS means dealing with additional hardware costs, employee onboarding costs and sometimes additional data collection. This can lead many potential GIS users to avoid the software before trying it in their context. Even when an office does start using GIS software, it can be difficult to realize value immediately. Like most B2B tools, one challenge in implementing GIS for real estate is user adoption within the company or government office. The tool can seem too complicated for anyone other than an analyst, so no one gets the full value.

Inconsistencies in data

Your decisions are only as good as the data you use to make them. Unfortunately, GIS tools often have inconsistent, inaccurate or outdated data. GIS software relies on spatial databases, which have many of the same challenges as any other type of dataset. Inconsistency stems from varied conceptualization and categorization and more technical issues like file management or data cleaning.

Lack of standardization

Because GIS tools have developed slowly over time, there is no heterogeneity between datasets and the resulting maps. There are very loose color conventions for geographic information, but adoption of these conventions is far from 100 percent. There is also quite a bit of variety in how elements are represented on the map: different icons for the same features, for example. And, finally, not every GIS tool has the same data layers. Real estate users have to assess which data layers they will need and ensure the tool has them before jumping in.

Self-assessment

Read the self-assessment guidelines for *Self-assessment scale for reading comprehension* and *Self-assessment scale for written expression* (based on the *China's Standards of English Language Ability*) outlined in the appendix of this textbook. Please conduct a self-assessment to evaluate your own skills respectively.

Unit 8 Sustainable Forestry

Learning Objectives

After learning this unit, you will be able to:

- Know about the sustainable forestry.
- Learn how to make inferences in academic reading.
- Design and structure a discussion-led academic writing.

Part I Reading for Academic

Purposes Making Inferences in Academic Reading

The capacity to make an inference is important in communication since language itself may be ambiguous and fragmented. In the academic reading process, making inferences requires readers to capture both the information presented in the text and to associate related information perceived previously. It is the cognitive process of generating new information and semantic meaning from or even beyond the text, which is an advanced requirement in reading comprehension. That is to say, to make inference, readers probably need to work along with two broad dimensions to combine semantic information of the text and non-semantic information related to the topic to achieve a thorough understanding or capture complete senses.

In foreign language reading tasks, the cognitive procedures of making inferences involve language decoding and information recoding. Language decoding refers to the generation of semantic information from foreign language forms. To decode an academic text, mastering knowledge of a foreign language (as units of meaning and form paring) is vital, especially in academic terminology, collocation pairs, sentence structures, cohesive and coherence devices, genres, and other language components. Common sense and academic knowledge, which have been acquired and accumulated before the reading task, are the basis of successful decoding and recoding in academic reading. It is beneficial to enhance inferencing capacities through learning to categorize topics, understand research background, theoretical framework, methodology, results, and identify the author's research interest in daily academic study.

In brief, making inferences in academic reading is essential for conceiving and making connection between both language in use (from genre to terminology) and scientific research contents (from theoretical attainment to research result illustration) to predict and infer adequate information through the text reading.

Reading Passage 1

 **Critical Thinking Questions**

Directions: Read the following passage and answer the following questions:

1. What are the main challenges in implementing sustainable forestry?
2. What is the writer's attitude toward the models and systems we use to make management decisions in the last paragraph?
3. How did the writer develop the main idea in the fourth paragraph?

Challenges in Implementing Sustainable Forestry

Philip J. Burton

A. Despite being an **enshrined** principle of responsible forestry for centuries, and even while the concept is being adopted by other sectors of society, it has been challenging to **implement** and demonstrate sustainable development in the context of operational forest management. Such challenges are understandable for a number of renewable resources—such as wild fish stocks that are difficult to track and annual agricultural crops that are sensitive to **vagaries** of a single season's weather—but why should it be difficult to manage forests sustainably, when trees and forests are long-lived, **stationary** and can be readily counted and measured?

B. Even if trying to simply achieve sustainable fibre (timber) production, one dimension of the problem is that the key elements of wood supply sustainability depend on estimation. Many key **parameters** and **coefficients** inherent to wood supply projections and allowable harvest determinations are fluid: estimates of growing stocks, growth rates and **regeneration** delays; and losses to pests, fires and storms are all subject to error and uncertainty. Year-to-year **variation** in weather conditions, seed crops and small mammal populations can affect regeneration success. Tree growth and mortality rates also vary with weather and weather events, and can depend on conditions of stand structure and inter-tree spacing that may be imperfectly characterized and understood. Timber losses to disturbance events—wildfires, windstorms, mass movements, floods, pest outbreaks—can be particularly difficult to anticipate; planning that depends on historical averages can grossly overestimate or underestimate impacts. With climate becoming more variable and disturbances more frequent under the effects of **anthropogenic** climate change, traditionally used forest yield models are becoming less reliable.

C. As if those biophysical challenges weren't enough, socio-economic and technological aspects of wood production present even more **intractable** challenges to sustainability. Oak woodlands historically **nurtured** to support shipbuilding lost value when ships were instead made of steel. The wisdom of planting or promoting one species or group of tree species over another is subject to the **whims** of market demand, as well as to the vagaries of species-specific outbreaks of native and invasive insect and **fungal** pests. Many forests reserved or planted for timber are now seen to have more value for amenity purposes such watershed protection, wildlife habitat and recreation, and so are protected from harvesting; this then requires adjustments to the rate of cut that can be sustained elsewhere in the forest estate. Growing human populations and **ballooning** real estate values mean that considerable forest land and timber production potential is lost to **exurban** sprawl and residential development, putting further pressure on the timber lands that remain. Globalized trading patterns are often able to offset regional differences in fibre supply and demand, but these are sometimes disrupted by **tariffs** and other trade barriers reflecting the politics of the day. Forest **interventions** (i.e. forestry and forest restoration) require financial investments, for which there are always alternatives: capital moves worldwide; land may be more valuable for agriculture or residential development; public funds are often diverted to health, education, infrastructure and defence. As forest products enterprises become increasingly concentrated in large, international corporations, investment capital is also increasingly mobile and **fickle**. Businesses undertaking sustainable wood production that protects the environment, have broad social support, and are economically **viable** can still be abandoned when profits aren't high enough.

D. The greatest challenge in implementing sustainable forestry is probably in achieving agreement on precisely what should be sustained, over what area of land, and with what priority to rank the forest values that inevitably conflict. <u>If diverse stand compositions and multiple canopy layers are desired to support **avian** biodiversity, for example, this may be **compatible** with non-motorized recreational activities, but at some cost to maximum conifer wood production.</u> Conversely, productive plantations may offer the best opportunity for rapid carbon sequestration, but may have little value for old-growth-dependent **vertebrates** and lichens. Sometimes a representative balance of these values can be sustained over a sufficiently large landscape through zoning and explicit resource-emphasis management in different zones. This still begs the question of how much land should be allocated to each resource emphasis (each value), and every such allocation decision is subject to our uncertainty as to "how much is enough". In general, the larger a land base available under a given management plan or policy regime, the better the prospects for **sustainability** of individual values and for the sustainability of net value to society. But large holdings often result in an unbalanced distribution of activities, so that some sites and some local activities bear the **brunt** of impacts that are clearly not sustainable at a local scale. For example, a village may face damaged vistas, fuelwood shortages and compromised wildlife habitat after rapid industrial logging in easy-to-reach traditional use areas, even though all those values may be sustainable over the geographic and temporal scope of the entire area being managed.

E. Forestry and its tradition of sustained yield also come with some undesirable baggage that is difficult to **jettison** or reform. Many forest management policies call for "maximum sustained yield" and an "even flow" of timber to sustain mills and forest-dependent employment and rural communities. Maximized yields demand management as close as possible to the feasibility frontier of decision-making, a frontier that is frequently overstepped when disturbances strike or the investment climate changes. The even-flow requirement may inspire a more conservative level of harvesting and assumptions about production, but ignores the event-driven sensitivity of forest ecology, business decision-making and politics. Sophisticated analysis of forest management systems has sometimes pointed out these vulnerabilities, but the response is often to attempt a greater level of command and control, a philosophy of natural resource management with innumerable failures and which inevitably inspires protests and **backlash** from local citizenry. The ultimate irony is that we see examples of forest practices that are in place to make forests follow the assumptions of the models and systems we use to make management decisions, rather than doing the hard work required to make our models and decision-support systems better match the real world.

Key Words and Phrases

1. enshrine /ɪn'ʃraɪn/ *v.* to make a law, right, etc. respected or official, especially by stating it in an important written document 把（法律、权利等）奉为神圣；把……庄严地载入
2. implement /'ɪmplɪment/ *v.* to make something that has been officially decided start to happen or be used 实施；执行
3. vagary /'veɪgəri/ *n.* unexpected and unpredictable changes in a situationor in someone's behavior that you have no control over 变化反复无常
4. stationary /'steɪʃənri/ *adj.* not moving; not intended to be moved; not changing in condition or quantity 不动的；静止的
5. parameter /pə'ræmɪtə(r)/ *n.* something that decides or limits the way in which something can be done 参数
6. coefficient /ˌkəʊɪ'fɪʃ(ə)nt/ *n.* a constant number that serves as a measure of some property or characteristic（数）系数；（物理）率
7. regeneration /rɪˌdʒenə'reɪʃn/ *n.* (biology) the process of growing again 再生
8. variation /ˌveəri'eɪʃ(ə)n/ *n.* variation (in/of something) a change, es-

		pecially in the amount or level of something 变化
9. anthropogenic /ænθrəupəu'dʒenik/	*adj.*	of or relating to the study of the origins and development of human beings 人为的；人类起源的
10. intractable /ɪn'træktəbl/	*adj.*	(of a problem or a person) very difficult to deal with 棘手的
11. nurture /'nɜːtʃə(r)/	*v.*	to care for and protect somebody/something while they are growing and developing 养育
12. whim /wɪm/	*n.*	a sudden wish to do or have something, especially when it is something unusual or unnecessary 突发的念头
13. fungal /'fʌŋgl/	*adj.*	of or caused by fungus 真菌的
14. balloon /bə'luːn/	*v.*	to increase rapidly in amount （数量上）猛增
15. exurban /ek'sɜːbən/	*adj.*	the outer suburb(s) of the city 城市远郊的
16. tariff /'tærɪf/	*n.*	a tax that is paid on goods coming into or going out of a country 关税
17. intervention /ˌɪntə'venʃn/	*n.*	action taken to improve or help a situation 干预
18. fickle /'fɪkl/	*adj.*	changing often and suddenly 善变的
19. viable /'vaɪəb(ə)l/	*adj.*	that can be done; capable of developing and surviving independently 可行的
20. avian /'eɪviən/	*adj.*	of or connected with birds 鸟（类）的；关于鸟（类）的
21. compatible /kəm'pætəbl/	*adj.*	able to exist or be used together without causing problems 可共存的；可共用的；兼容的
22. vertebrate /'vɜːtɪbrət/	*n.*	any animal with a backbone, including all mammals, birds, fish, reptiles and amphibians 脊椎动物
23. sustainability /səˌsteɪnə'bɪləti/	*n.*	the use of natural products and energy in a way that does not harm the environment; the ability to continue or be continued for a long time 持续性
24. brunt /brʌnt/	*n.*	usually in phrases, such as to bear the brunt or take the brunt of something unpleasant means to suffer the main part or force of it 受主要冲击

25. jettison /'dʒetɪs(ə)n/ *v.* to get rid of something/somebody that you no longer need or want 丢弃；处理掉

26. backlash /'bæklæʃ/ *n.* a strong negative reaction by a large number of people, for example to something that has recently changed in society（对政治或社会变化的）强烈反应

Reading Comprehension

Exercise 1

Directions: Read the **Passage 1** again and answer the questions below. Please give brief answers within 10 words.

1. According to the text, how to solve the the greatest challenge in fulfilling sustainable forestry?

2. For a number of renewable resources, why is it difficult to manage forests sustainably?

3. Besides the market value of woodlands, what other social values are found?

Exercise 2

Directions: Read the **Passage 1** again and decide if the following statements are true or false. Circle T for true and F for false.

1. (　　) Renewable resources—such as wild fish stocks are dynamic and differ from long-lived, stationary, trees and forests.

2. (　　) Many forest management policies are highly related to sustain local employment and rural communities.

3. (　　) Sharing agreement on the facts of sustained forest is the primary issue to deal with.

4. (　　) Social value of forest is only driven by the market demand in this 21st century.

5. (　　) Forest yield model was popular to estimate forest sustainability

Vocabulary Exercises

Exercise 1

Directions: In this section, there are ten sentences, each one with one word missing. You are required to complete these sentences with the proper form of the words given in the brackets.

1. The contract ____________ that a penalty fee is to be paid if the work is late. (specific)

2. To be ____________, they have to think about the future and manage the waste and the sewage water. (sustain)

3. As is said by many, you can live in heaven or hell, it is your choice and your

____________. (perceive)

4. In many cultures, temples are located in mountains to ____________ giant status of god or Buddha. (enshrine)

5. Investigating the biological____________ of different materials is important in land protection. (compatible)

6. The ____________ of plant gene can influence its growth circle. (vary)

7. Predictability is one of the____________ that reduces or eliminates the negative effects of noise. (vary)

8. Environmental____________ reflects the policy, market demand and trading patterns. (disrupt)

9. Biological treatments could be developed into long-term ____________ to impact the structure of the brain. (intervene)

10. Brain cells have limited ability to ____________ if destroyed. (regeneration)

Exercise 2

Directions: Match the words with their definitions.

____1. hazard	A. an act or instance of burning
____2. reclamation	B. the attempt to make land suitable for building or farming
____3. deplete	C. the act or process or an instance of leaking
____4. eradication	D. something that is dangerous and likely to cause damage
____5. combustion	E. the complete destruction of something
____6. arid	F. to make a person or a group of people unimportant in an unfair
____7. leakage	G. more advanced or complex than others
____8. marginalized	H. to fail to guess or understand the real cost, size, etc. of something
____9. underestimate	I. to reduce something in size or amount
____10. sophisticated	J. lacking sufficient water or rainfall

Exercise 3

Directions: In this section, there are ten sentences with ten blanks. You are required to select one word for each blank from a list of choices given in a word bank. Each choice in the bank is identified by a letter. You may not use any of the words in the bank more than once.

A. accelerate	B. accumulate	C. alternative	D. channeled
E. distinguish	F. implemented	G. irreversible	H. launched
I. minimize	J. sustainable	K. restored	L. yield

1. Money for the project will be ____________ through the local government.

2. They ____________ a big advertising campaign to promote our new toothpaste.

3. People with self-disciplined can not only ____________ wealth but also establish

their careers.

4. The best you can hope for is to ____________ the damage that can occur.

5. Agricultural scientists are increasingly aware of slow, ____________ trends in soil and climate deterioration.

6. They used chemical treatments to ____________ the growth of crops.

7. The changes to the national park system will be ____________ next year.

8. The same seed and fertilizer program may ____________ completely different outcomes in different places.

9. Seeking ____________ sources of nutrition to animal meat is one of the solutions to reduce carbon dioxide emissions.

10. In answering this question, it is essential to ____________ between aspects of market demand.

Exercise 4

Directions: Read the passage again and translate the following sentences from **Passage 1** (underlined in the passage) into Chinese.

1. Many key parameters and coefficients inherent to wood supply projections and allowable harvest determinations are fluid: estimates of growing stocks, growth rates and regeneration delays; and losses to pests, fires and storms are all subject to error and uncertainty.

2. If diverse stand compositions and multiple canopy layers are desired to support avian biodiversity, for example, this may be compatible with non-motorized recreational activities, but at some cost to maximum conifer wood production.

3. Maximized yields demand management as close as possible to the feasibility frontier of decision-making, a frontier that is frequently overstepped when disturbances strike or the investment climate changes.

Part II Academic Writing Strategies

Argumentation/ Reasoning: Defending a Counter Voice

Argumentation is one of the most common genres in academic writing at the tertiary level. Presenting a clear position and justifying it on a particular issue or question is vital in introducing an argument. A discussion-led argument is one of these approaches to the issue by presenting opposing views before defending your statement on the issue. In this approach, three steps are worth to be pointed out to arrange your writing.

The first step is to provide an appropriate context or lead-in to the issue to prepare readers for a specific issue or topic. It requires presenting information on opposing views. The opposing views can be common sense or citations of studies (direct or indirect). It is the key to preparing readers to understand your following persuasion and reasoning. In this step, explaining the issue prepares readers with a clear understanding with what is the specific issue and why there is disagreement.

The second step is to justify your position and statement with scientific findings and examples in response to opposing views. In developing the argument in the body of the essay, two core questions should be borne in mind: what is your thesis or argument and how could the selected material support your statement? Justifying a discussion-led argument is not simply stated with the comment "I disagree", but rather to set up the following text to support your position and statement by reasoning and evidence. Previous research findings, examples, data evidence, scientific statements, and logical analysis are helpful to justify and defend your statement. To indicate your counter-voice in academic writing, evidence and examples should be relevant to your analysis, interpretation, and evaluation.

A conclusion, as the third step, is necessary. For example, comparing the opposing view with your statement or reviewing the issue from your viewpoint, helps reinforce the reader's understanding or acceptance of your statement on the issue.

Reading Passage 2

Critical Thinking Questions

Directions: Read the following passage and answer the questions:

1. What is the format of the on-line paper?
2. What is the main idea of this passage?
3. How does the writer organize the introduction?
4. What is the relationship between the sustainable development and sustainable forest development?

Introduction of Sustainable Forest Development

Olga Ikonnikova, Aleksandr Gorkin, Vitaliy Petrik

A. Achieving stable economic growth was one of the main goals of any state in the 20th century. The thoughtless **striving** at any cost to demonstrate to the world the growth of the gross domestic product, not supported by the introduction of new technologies and care for the environment, almost led to a global collapse. Suddenly it turned out that nature is not able to meet the growing needs of humankind.

B. In 1972, the Roman Club, an international public organization that made a significant contribution to the promotion of the idea of **harmonization** of human relations with nature, presented the famous "Limits of Growth" report, prepared by Donella and Dennis Meadows and their colleagues, which aroused enormous public **resonance** and for the first time so clearly and loudly **denoted** the problem of the critical state of the **biosphere** as a complete self-organizing system. Scientists were able to show that, while maintaining current trends in increasing production and the population of the planet, the limits of growth on Earth can be achieved in the near future. And although the **scenarios** of the development of the situation predicted by them were largely based on **erroneous** assumptions, the value of this report is unquestionable.

C. In 1972, the United Nations Conference on the Human Environment was held and the United Nations Environment Program was established, which indicated that the world community was already connected to the solution of this problem at the state level.

D. One of the first international documents that referred to the term "**sustainable** development" was the World **Conservation** Strategy, developed by the International Union for Conservation of Nature and Natural Resources, the UN Committee on the Environment and the World Wildlife Fund, which emphasized the need to take environmental factors into account Social and economic development.

E. In 1983, under the **auspices** of the United Nations, the World Commission on Environment and Development was established, headed by Gro Harlem Brundtland, whose name is now associated with the **dominant** interpretation of the term "sustainable development". In 1987, the Commission published the report "Our Common Future", in which sustainable development was defined as development, in which meeting the needs of the present generation does not undermine the ability of future generations to meet their own needs.

F. In 1992, the UN Conference in Rio de Janeiro adopted the Declaration on Environment and Development, one of the principles of which is to ensure **equitable** satisfaction of the needs of present and future generations. Central to the efforts to achieve sustainable development in accordance with the Declaration is the concern for people who have the right to a healthy and fruitful life in harmony with nature. An integral part of the development process should be the protection of the environment.

G. In 2002, at the World Summit on Sustainable Development held in South Africa, it was concluded that the world had not followed the path of sustainable development for two decades. The summit participants **reaffirmed** their commitment to Agenda 21 and the Declaration adopted in Rio de Janeiro, but the recommendations adopted at the summit, basically, remained on paper. Most environmental, economic, social and political problems continue to worsen. A similar conclusion was reached in 2012.

H. One should not think that the problems of environmental protection and the development of the principles of sustainable development were of concern only to foreign scientists. Back in 1964, a geographic edition of the Mysl Publishing House published a book by Soviet geographer and ecologist David L. Armand "*To Us and Grandchildren*," in which it was noted that "it is the moral duty of each generation to leave the next natural wealth in better condition and in greater quantity than it got from the previous one."

I. One can say that the **predecessor** of the concept of sustainable development was the "concept of the **noosphere**", expressed by the Russian scientist V.I. Vernadsky, in which the noosphere is defined as the highest stage of the evolution of the biosphere, by which humanity is transformed into a new **geological** force, transforming the face of the planet with its thought and labor.

J. The concept of sustainable development unites three main components: economic, social and environmental. The economic component is based on the theory of the maximum flow of **aggregate** income; the social component focuses on preserving the stability of social and cultural systems, while the ecological component seeks to ensure the **integrity** of biological and physical natural systems.

K. Thus, sustainable development is possible only on the basis of harmony of relations among man, society and nature.

Key Words and Phrases

1. strive /straɪv/	*v.*	the act of trying very hard to achieve something 努力
2. harmonization /ˌhɑːmənaɪ'zeɪʃn/	*n.*	the quality of two or more things going well together and producing an attractive result 和谐化
3. resonance /'rezənəns/	*n.*	(in a piece of writing, music, etc.) the power to bring images, feelings, etc. into the mind of the person reading or listening; the images, etc. produced in this way 共鸣
4. denote /dɪ'nəʊt/	*v.*	to be a sign of something 显示
5. biosphere /'baɪəʊsfiə(r)/	*n.*	the part of the earth's surface and atmosphere in which plants and animals can live 生物圈
6. scenario /sə'nɑːriəʊ/	*n.*	a description of how things might happen in the future 设想
7. erroneous /ɪ'rəʊniəs/	*adj.*	not correct; based on wrong information 错误的
8. sustainable /sə'steɪnəb(ə)l/	*adj.*	that can continue or be continued for a long time 可持续的
9. conservation /ˌkɒnsə'veɪʃ(ə)n/	*n.*	the protection of the natural environment（对环境的）保护
10. auspice /'ɔːspɪs/	*n.*	the help, support or protection 支持
11. dominant /'dɒmɪnənt/	*adj.*	more important, powerful or easy to notice than other things 处于支配地位的

12. equitable /'ekwɪtəb(ə)l/ *adj.* fair and reasonable; treating everyone in an equal way 公平合理的

13. reaffirm /ˌriːə'fɜːm/ *v.* to state something again in order to emphasize that it is still true 重申

14. predecessor /'priːdəsesə(r)/ *n.* a person who did a job before somebody else 前辈

15. noosphere /'nəʊsfɪə/ *n.* the part of the biosphere that is affected by human thought, culture, and knowledge 心智层

16. geological /ˌdʒiːə'lɒdʒɪkl/ *adj.* connected with the scientific study of the physical structure of the earth, including the origin and history of the rocks and soil of which the earth is made 地质学的

17. aggregate /'ægrɪgət/ *n.* a total number or amount made up of smaller amounts that are collected together 合计

18. integrity /ɪn'tegrəti/ *n.* the state of being whole and not divided 完整

Writing Tasks

Exercise 1

Directions: Read the following sentences and rewrite the following sentences (underlined in the passage), replacing the underlined parts with your own words.

1. And although the scenarios of the development of the situation predicted by them were largely based on erroneous assumptions, the value of this report is unquestionable.

2. In 1983, under the auspices of the United Nations, the World Commission on Environment and Development was established.

3. In 1992, the UN Conference in Rio de Janeiro adopted the Declaration on Environment and Development, one of the principles of which is to ensure equitable satisfaction of the needs of present and future generations.

Exercise 2

Directions: Read the **Passage 2** again and answer the questions below. Please give brief answers in about 10 words.

1. What problem did the "Limits of Growth" report indicate for the first time? (Para. B)

2. Why did the Russian scientist define "noosphere" as the highest stage of the evolution of the biosphere? (Para. I)

3. In sustainable development, how should economic, social and environmental components balance with each other? (Para. J)

Exercise 3

Directions: Read the passage again and write a summary of **Passage 2** with no more than 180 words.

Extensive Reading and Writing

Directions: For this part, you are required to read the following passage and then write a summary of the author's idea followed by your statement with supportive materials. You should write at least 120 words but no more than 180 words.

Traditional Forestry, Sustainable Forestry and Forestry Sustainability

Lucio Muñoz

Forestry can be defined as the set of practical and theoretical concepts and tools that permit the management of forested and deforested areas in accordance to dominant views of development. As new dominant views of development replace old ones, the internal structure or goals of the forestry model change reflecting new values and social attitudes. Below there is a general description of how forestry programs have evolved from traditional economic roots alone to its green form and to the possible evolution towards a sustainability form.

1. Traditional forestry

It is known that traditionally forestry was used and promoted as a source of economic values only. Pearce points out that forest policy was decided until very recently on the economic value of timber alone as the value of non-timber products and services, with markets and without markets, was left out of the model. Hence, forestry programs and plans were directed at the creation of value through the transformation of existing forested areas into timber commodities or through the conversion of existing deforested areas into plantations. As such, this was the economic value driven forestry model where non-timber values and non-economic values were not considered relevant when planning the management of forested and deforested lands. As non-timber issues became relevant, a shift in the design and implementations of forestry programs and practices started to take place. The Centre for International Forestry Research reports that it took increased interest on the role of non-timber forest products in 1998 as a potential tool for planning effective development and conservation programs and it began major global efforts to research and understand better the role non-forest products issues in development.

2. The greening of traditional forestry

The increasing importance of environmental concerns and the need to include ecological values within forestry programs has led to the redefinition of these programs and their tools so as to make them the principal components of a new eco-economic engine, a system capable of creating both economic and ecological values at the same time.

Hence, there is a need now to adjust traditional forestry programs to incorporate non-timber products and services issues too; and discussion, research and action on how to best do this is currently underway. For example, it became relevant now the need to understand the role of and to assess the size of non-timber values to support management/land use decisions; the goals of protection and sustainable used of forests in a balanced way became central concept to forestry strategies of institutions like the World Bank; a survey of different approaches to deal with payment for environmental services implemented in the western hemisphere has been carried out to determine best practices and conditions for success as well as approaches aimed at creating markets for biodiversity had been discussed and proposed; and Interconnecting issues such as climate change and forest are now of central concern to existing forestry commissions such as the North American Forestry Commission. The forestry model in which ecological concerns are paired with economic concerns is the one now known as the sustainable forestry model, and which the author calls the eco-economic forestry model.

3. Evolving forestry practices

Recently, another forestry concern has been added to the discussion: the need to incorporate social issues within the eco-economic forestry model. For example, the growing need to add social interactions to thc cnvironment-economy connection is behind the elaboration of the reference manual for the integrated assessment of trade-related policies available now to both developed and developing countries alike for consideration and use; the use of sustainable resource management is seen now as a key tool to deal with poverty issues; and one of the key changes to the development as usual model that need to be made to attain a sustainable economy is the inclusion of social/poverty issues. The need to account for social concerns is based on the premise that the incorporation of economic and ecological values is a necessary, but not a sufficient condition for sustainability to take place, including forestry sustainability. Hence, the evolution of forestry practices is driven by the substitution of old value systems by new ones, and this evolution appears to be heading towards forestry sustainability. In other words, the need to include social issues will push sustainable forestry towards sustainability forestry when they become binding. The United Nations Environmental Program Year Book for 2009 proposes in Chapter 1 the need to manage ecosystems in ways that efficiently deal with the poverty issue, hence making the incorporations of social concerns in development essential to sustainability.

Self-assessment

Read the self-assessment guidelines for *Self-assessment scale for reading comprehension* and *Self-assessment scale for written expression* (based on the *China's Standards of English Language Ability*) outlined in the appendix of this textbook. Please conduct a self-assessment to evaluate your own skills respectively.

参 考 文 献

References

丁往道，吴冰，2011．英语写作基础教程［M］．3 版．北京：高等教育出版社．

季佩英，范烨，2013．学术英语［M］．综合版．北京：外语教学与研究出版社．

约翰·兰甘，佐伊·奥尔布赖特，2014．美国大学英语写作［M］．9 版．北京：外语教学与研究出版社．

张玉娟，陈春田，邹云敏，等，2019．新世纪实用英语写作［M］．北京：外语教学与研究出版社．

张在新，2011．英语写作教程［M］．北京：外语教学与研究出版社．

中华人民共和国教育部，国家语言文字工作委员会，2018．国家语言文字规范：GF0018－2018 中国英语能力等级量表［S］．北京：高等教育出版社．

Bailey, Stephen, 2015. Academic Writing for International Students of Business [M]. London: Routledge.

Chikalanga, Israel, 1992. A suggested taxonomy of inferences for the reading teacher [J]. Reading in a Foreign Language, 8: 697-709.

Margaret Cargill, Patrick O'Connor, 2013. Writing Scientific Research Articles [M].2nd ed. New Jersey: Wiley-Blackwell.

Last, Susan, 2024. Technical Writing Essentials [M/OL]. [S.l.] Pressbooks. https://pressbooks.bccampus.ca/technicalwriting/front-matter/introduction/.

Richard Thurlow, Paul van den Broek, 1997. Automaticity and inference generation during reading comprehension [J]. Reading & Writing Quarterly, 13 (2): 165-181.

Read, Siew Hean, 2019. Academic Writing Skills for International Students [M]. London: Red Globe Press.

注：为了方便读者通顺阅读，对于原论文中引注的参考文献的标注进行了删除，特此声明。

附　　录

Appendix

附表 1　阅读理解能力自我评价量表（Self-assessment Scale for Reading Comprehension）

级别	能力描述	自我评测（Y for yes or N for no）
九级	● 我能读懂语言复杂、内容深刻、跨专业的各类材料，如学术论文、文学原著或应用文等。 ● 我能从多个角度评价阅读材料的价值	
八级	● 我能读懂语言复杂的文学原著，并欣赏作者的语言表达艺术。 ● 我能读懂专业领域相关的学术论文或科技文献并评价其研究方法。 ● 我能通过研读多篇同类型阅读材料，综合评价作品的语言风格	
七级	● 我能读懂语言复杂的小说及文化类作品，并鉴赏作者的语言特点。 ● 我能读懂相关专业领域的书评。 ● 我能通过浏览目录，预测全书（文）的主要内容	
六级	● 我能读懂语言较复杂的小说和议论文。 ● 我能理解和概括说明性材料中被说明对象的主要特征。 ● 我能读懂相关专业领域的操作指令并理解专业术语。 ● 我能通过阅读材料的选词、修辞方式等，推测作者态度	
五级	● 我能阅读一般题材的议论文或话题熟悉的评论性文章。 ● 我能理解语言较复杂的社会生活故事中的各种常见修辞手法。 ● 我能在读应用文，如会议纪要等时，提取主要信息。 ● 我能在阅读中适时概括已读过的内容	
四级	● 我能阅读简短的故事、散文或说明文。 ● 我能读懂旅游见闻中关于事件、人物、地点等信息。 ● 我能从社会生活相关的简短议论文中分析作者的观点。 ● 我能利用略读、寻读、跳读等不同的阅读技巧，找出文章中的重要信息	
三级	● 我能读懂语言简单的各类故事。 ● 我能从散文中提取人物、景物及其他细节信息。 ● 我能理解简短书信中作者的观点。 ● 我能通过关键词或主题句理解文章的主要内容	
二级	● 我能读懂语言简单的有关日常生活的短文。 ● 我能从便条、通知、任务指令等材料中获取具体信息。 ● 我能从文章的描写中概括人物或事物的主要特征。 ● 我能借助插图理解图文小故事	
一级	● 我能阅读语言简单的绘本或小故事，识读其中的常见词并理解其主要内容。 ● 我能感受童谣中的押韵	

附表 2　书面表达能力自我评价量表（Self-assessment Scale for Written Expression）

级别	能力描述	自我评测（Scale from 1—5，low to high）				
		1	2	3	4	5
八级	● 我能对所收集的文献进行综述和评价。 ● 我能撰写学术会议发言稿					
七级	● 我能就专业相关话题整理文献。 ● 我在听学术讲座时，能做准确、详细的笔记。 ● 我能在完成论文初稿后从内容、结构和格式等方面进行修改					
六级	● 我能撰写专业论文的英语摘要。 ● 我能用同义词或近义词来避免表达中的词语重复					
五级	● 我能写国际交流项目的申请书。 ● 我能根据老师或同学的反馈修改文章结构和内容。 ● 我能写符合学术规范的议论短文					
四级	● 我能根据所读材料写概要。 ● 我能在写作前列出写作提纲。 ● 我能用主题句突出段落的主旨大意。 ● 我能检查并改正作文中的用词和衔接错误					
三级	● 我能在写作文前收集有用的词句。 ● 我能使用连接词来连接句子。 ● 我能检查并修改明显的语法错误